Mustard Doesn't Go on Corn!

How respect, openness, and a simple process for
innovation can lead to great ideas

Richard Trombetta

Order this book online at www.trafford.com/05-2897
or email orders@trafford.com

Most Trafford titles are also available at major online book retailers.

Note for Librarians: A cataloguing record for this book is available from Library
and Archives Canada at www.collectionscanada.ca/amicus/index-e.html

Printed in Victoria, BC, Canada.

ISBN: 978-1-4120-7999-0

*We at Trafford believe that it is the responsibility of us all, as both individuals
and corporations, to make choices that are environmentally and socially sound.
You, in turn, are supporting this responsible conduct each time you purchase a
Trafford book, or make use of our publishing services. To find out how you are
helping, please visit www.trafford.com/responsiblepublishing.html*

*Our mission is to efficiently provide the world's finest, most comprehensive
book publishing service, enabling every author to experience success.
To find out how to publish your book, your way, and have it available
worldwide, visit us online at www.trafford.com/10510*

www.trafford.com

North America & international
toll-free: 1 888 232 4444 (USA & Canada)
phone: 250 383 6864 ♦ fax: 250 383 6804
email: info@trafford.com

The United Kingdom & Europe
phone: +44 (0)1865 487 395 ♦ local rate: 0845 230 9601
facsimile: +44 (0)1865 481 507 ♦ email: info.uk@trafford.com

10 9 8

I would like to dedicate this book to my wife who constantly encourages me to carry out my mission and to my daughter who never ceases to teach me new and exciting things every day.

I would also like to dedicate this book to my parents, who continue to provide unconditional support despite their struggle to come up with an answer to the question 'what does Rich do?'

Special acknowledgements go to the following people who provided great feedback and encouragement along the way. In alphabetical order they are: Richard Bitner, Tim Burke, Andy Cheng, Richard Eaton, Lynne Eickholt, Faithe Hart, Judy Johnston, Carrie Kuempel, Christie Mabry, Lisa O'Connell, Patricia Piquette, Richard Hilbert, Trevor Snorek-Yates, Chris Trombetta, Charlotte Tyson, Tracy Wheeler, Gary Zukowski

Table of Contents

Introduction

Introduction

Why is this book being written?

T his book is not just a book. It is a calling. It is my mission. It is the result of over 15 years of working in corporations – large and small – and continuously being frustrated with the approach many organizations take to promote innovation and creativity.

The flow and style of this book is designed to create an enjoyable, fun, and practical experience for the reader. One that is fluid, entertaining, educational, and above all, *helps you increase your bottom line and achieve your business objectives*. My goal is to inspire you. Regardless of your position in your company, my goal is to inspire you to join a movement in corporate America that will quickly and easily enable all employees to constantly be sharing and implementing new ideas.

In my mind I see two people reading this book at the exact same moment. The first is the boss of a group of people. The second is a person who reports to that boss. They are unaware that the other is reading this book. They both finish the book, jump up from their chairs, race down the hall, and collide. Simultaneously they both yell out, "You have to read the book I just finished! It will change our company!" Together, they will do so. Together they will tell others about how innovation is easy and everyone is creative. Profits will rise, customers' expectations will be exceeded, and employees will love their jobs. And the word will spread. Soon it will not just be their department or team, but their entire company. And then their company's clients and business partners. And so on and so on. There will be an explosion of ideas so loud it deafens phrases like 'the problem with that is' or 'that won't work because' and replaces them with respect, openness, and a simple process for innovation. Paradigms will be smashed and for many, it will raise their confidence levels to new heights. This book will be the start of this process.

I must stress that this is not a doctoral thesis or a book grounded in extensive research. I wish I had both the patience and academic rigor to put Ph.D. after my name, but I do not. Regardless, it would not fit with the theme of the book – innovation is EASY. With that in mind, this book will be easy to read, easy to use, and easy to buy in mass quantities. Wait, did I say that out loud?

Many authors begin with a book and then follow-up with a field guide. I am going right to the field guide. A friend of mine once said, 'this is the decade of the practical.' I agree and that is what I am trying to create – something you can read and immediately put into action. There will be stories, movie and TV examples, and even worksheets. I will also provide some simple things you can try with co-workers. My experience is that people can often relate better to a new concept or idea when they can 'see' it. Recognizing this, I have included numerous stories and examples to help the words come off the page and paint a picture in your mind. A picture says a thousand words, right? Just something to think about next time you pick up a text book.

Chapter 1 overviews why it is difficult for companies to be innovative. This section provides the groundwork for Chapter 2, which focuses on the simple steps companies can take to become truly innovative. Chapter 3 reviews some fundamental topics such as prioritization and personality differences and Chapter 4 provides an outline of what you need to do to get the approach in this book moving in your company.

Note: All of the action items and key points in this book are also located at www.innovationiseasy.com/actionitems.htm

But why is the book titled "Mustard Doesn't Go on Corn!?"

I once brought my young daughter to a small children's museum. Everywhere I looked I saw words like explore, discover, and imagine. In one section of the museum there is a small play kitchen that can accommodate about 10 kids. I was watching my daughter have a grand old time putting plastic grapes in the play oven when I saw a remarkable event. There was a little boy about 3 or 4 years old who had a plate with some plastic corn on it. He said to his mom, "OK, mom, I'm going to put mustard on your corn." Just as he was about to do so his mom said in a semi-nurturing voice, "mustard doesn't go on corn." It was at that moment in time I realized why innovation is so difficult for companies and our society.

Here was a little kid seeing words and images encouraging him to explore and imagine and the second he does, boom –'mustard doesn't go on corn!' The kid's face dropped. What made it worse is what happened next. Another little kid very emphatically said, "No, mustard doesn't go on corn." You may be asking 'what's the big deal here?' But just think, in a matter of seconds the kid had his idea shot down by an authority figure and was piled on by a peer. Yikes! Being the instigator I am, I could not just sit back and watch this happen. I said, "I like mustard on corn." The kid looked confused. "I do. I put mustard on everything I eat." His eyes got big and he smiled. "You do?" he asked. "Yup, even on spaghetti." Suddenly energy started to take over that little kitchen. Within seconds other kids were getting involved and now mustard on corn didn't seem so foolish after all.

Let's play out this same scenario at work. A company has an 'innovation initiative' and puts up signs and banners with phrases like 'every idea counts' or 'innovation is king.' A person suggests an idea. The boss publicly says 'that won't work.' A co-worker then says, 'yeah that's not a good idea.' Ka-boom! That person's world is crashing in around him. A couple of days later in a meeting the boss asks for ideas. The person who had his idea slammed sits there quietly. He says nothing. After the meeting a co-worker asks him why he didn't share any ideas. He simply says, "I'm not creative."

Do you want to join me in ending stories like this? Do you want to transform your company? Do you want to greatly exceed your business objectives and the needs of the marketplace? If you do, please join me on my mission.

Chapter One

Why is becoming an innovative company so difficult?

"One night I was laying down and I heard Momma and Poppa talking. I heard Poppa tell Momma 'let that boy boogie-woogie. It's in him and it's got to come out.'" - Blues legend John Lee Hooker

ere is my mission:

To inspire individuals and organizations to create respectful, open, and innovative work environments that promote the creation of new ideas.

While I have participated in several useful workshops on innovation and there are good books on the subject, the concepts and tools can be complicated. My philosophy is very simple. Create a culture in which all employees are open to ideas AND respect others for their input, and innovation will flourish. Yes, it's that easy. I call this a POP! Culture® - applying the Positive Outlook Principle to achieve innovation. When companies, groups, or teams have a POP! Culture, openness and respect dominate the work environment and there is an explosion of ideas. You may be asking, 'but what will we do with all of these ideas? It will be overwhelming.' That is where my process for innovation, NEWIDEA!™ comes in. Companies that create a POP! Culture and support it

Innovation = respect and being OPEN to ideas.

Simple

with the NEWIDEA! process I outline in this book have the potential to be successful beyond their wildest imaginations. Employees will not only be constantly sharing ideas, they will be constantly implementing them as well. To learn more about a POP! Culture visit www.innovationiseasy.com

Why respect and openness?

Howard Baldwin, in the January 2005 edition of *Optimize Magazine,* writes "Polls by the Gallup Organization have shown that while 25% of the work force is engaged in their work, 60% are disengaged, and the other 15% are actively disengaged. There's a direct bottom-line impact to this. Again going back to the Gallup polls, business units with employee engagement scores in the top half of rankings have 86% higher customer-satisfaction ratings, 70% more success in lowering turnover, 44% higher profitability, and 78% better safety records."

Question:
Are your employees truly engaged in their work?

If someone feels they are not being respected or that people are not open to his or her ideas, this person will begin to drift into the disengaged category. While the definition of disengaged may be open to interpretation, here is an illustration that is not debatable. If you have employees who consciously choose not to share suggestions because they feel their ideas are not respected or that the environment they work in is not open to their ideas, then these people are disengaged. *They eventually fall into the actively disengaged category and negatively impact your bottom line.*

Story: Free consulting

A colleague of mine shared a story with me that illustrates the impact that a disengaged employee can have on innovation. He had been invited by a consulting firm to participate in a problem solving session with a group of facilitators who had been recently trained in a process around creativity. To maximize the benefits of the consulting time, he was allowed to hand pick eight individuals to join him as participants in the discussion. "At one point in the brainstorming process I threw out an idea," he told me. "All I heard were a few 'sniffs.' Here we were in a room with people I hand picked to help me, with trained facilitators stressing not to judge ideas and people just 'sniffed.' I can recall having at least ten other ideas that I did not share during that meeting. I remember thinking of an idea and then making a conscious decision not to share it because of the reactions of the group."

You may be thinking, 'so he wants us to act on every idea and never pass any judgment?' No! Here are points I can't stress enough.
- Not every idea gets acted upon
- This is much more than people just saying 'yes' or 'yes and!' to ideas
- Not all ideas justify often scarce and limited resources
- Healthy debate around ideas is good

It would be irresponsible and a complete waste of time for companies to have an environment where the expectation that was every idea would be acted upon. In addition, without healthy debate organizations would be taking action on ideas without acknowledging and appreciating diverse points of view.

My approach is not getting everyone to feel "happy, happy, joy, joy." You are in business. Businesses have objectives and financial commitments. This book will help you reach them AND create outstanding work environments. I have spent over 15 years working for great companies like GE, Fidelity Investments, and Thomson Financial and I understand the demands corporations put on individuals. In addition, as an electrical engineer, I have been presented with many complex challenges and situations that have required both practical academic discipline and creative approaches to solving problems. After 15 years in corporate America and seeing and experiencing the struggles that many organizations have with innovation, I had a revelation. Innovation is very, very simple. It is one of the easiest things for a company to do because it is simply about creating a culture in which everyone is open to ideas and respects others for their input. Openness and respect. Openness and respect. It sounds simple - and it is. Unfortunately, American society and corporate cultures have taken something so easy and made it extremely difficult.

This approach is not 'happy, happy, joy, joy.'

If you were sitting down with me right now, I would expect you to ask me three questions:

1. How do I define innovation?
2. Can I prove respect and openness are the keys to innovation?
3. This is too simple. How can innovation be this easy?

1. How do I define innovation?

Here is my definition - Innovation is the creation of and action upon ideas to develop a new process, concept, or material object that solves a problem or addresses an apparent or latent need in the marketplace.

Please notice the words 'action upon.' An organization can have all the respect and openness in the world, but if it is not *doing something* with the ideas that come as a result of their culture, is innovation really present? I will discuss my process for turning ideas into action, NEWIDEA!, later in the book.

2. Can I prove respect and openness are the keys to innovation?

While I may not have all the scientific evidence or years of extensive research to withstand the scrutiny of skeptics, I will rely on basic business principles and common sense. Here are just a few of the tangible benefits you could see as a result of implementing the concepts I propose:

Ideas without action are just...ideas.

7

Benefit	Potential impact to your bottom line
More productive discussions in meetings and increased productivity	If everyone reduces time spent in meetings by only 1 hour per week, you gain an entire week (50 hours) of productivity from each person each year!
Higher retention rates	Multiply 1.5 times the salary of any person who leaves and shares statements such as 'they don't value my input'
More satisfied customers	It is 10 times more expensive to replace a customer than to keep an existing customer
More new solutions being created	As an example, a simple idea like Post-it Notes© by 3M can generate millions of dollars of income each year. Are you missing simple ideas in your company? What is this costing you?

Also, going back to common sense, by having a culture in place that allows new ideas and innovation to flourish, the opportunity for dramatic effects on your bottom line increases as well. The two are directly proportional. The more ideas, the better your chance of success.

3. This is too simple. How can innovation be this easy?

This is the best part of what I am proposing. It is simple. It is easy. I once had an engineering professor share something with me that I have never forgotten. He told me, "If you are struggling to solve a problem on an exam, immediately think of the most obvious and simple solution." I was confused. This went against conventional thinking that exams would be difficult and challenging. "Think of it this way," he said. "I have to correct over 100 exams, each containing page after page of mathematics and engineering principles. I design the answers to be things such as 'zero' or 'two times Pi', not some complex value that would create headaches for me while grading a student's work. The problem may appear complex but the answer is usually quite simple." The same can be said for innovation. What may appear to be complex actually has a simple answer, and it is right here in this book.

If you took a glass and smashed it against a wall, that is what I am trying to do with innovation. Take all of the complex charts and theories and smash them. Go for the simple. My goal is to blaze a new trail for companies to travel to better achieve their objectives. Welcome aboard.

The Balloon Effect™

While there are many steps in the creativity process that could be explored in great detail, it is the *split second* when an idea or suggestion is shared that I am primarily focused on in this book. It is this moment in time when an idea will take one of two paths. The idea will die or the idea will have a chance to grow. It is that straightforward.

To illustrate this point, I would like to share with you what I call The Balloon Effect.™ In my POP! Culture workshops I conduct an exercise that you may want to try with your team. I ask half the participants to begin blowing up balloons. As they blow up the balloons I tell them that what they are doing is similar to sharing an idea. It takes energy. It is visible. As the energy grows, so does the idea. I then ask the remaining participants to take a paper clip, unbend it, and form a pin. You can probably guess where this is going. I tell the people with the pins that they are going to pop the balloons. But not just simply pop them. I tell them that when they pop the balloons they need to look the person whose balloon they are popping right in the eyes and say 'the problem with your idea is' or 'that won't work because' right before they pop it. Pop go the balloons!

You may be saying, 'okay, I get it. Don't pop people's balloons.' But that is not the real lesson here. What I then ask the group is, 'Now that you have seen what has happened, do we even need the pins or will just the

> It is the *split second* an idea is shared that determines the path the idea will follow.

10

thought of someone popping your balloon be enough to discourage you from doing this exercise over and over again?' My point is that in companies, once people figure out 'pins' are present, people stop blowing up balloons (i.e. sharing ideas). And let's take it a step further. Let's say you are new to a company and you see someone's idea popped in a meeting in a very disrespectful manner. How willing are you going to be to share a new idea? Or what if you are a shy, introverted person? So, the point is not about popping balloons – it is that balloons stop being blown up and ideas don't even come out!

Action Item #1:	Share the Balloon Effect concept with a co-worker and get his or her response.

Person I will try this with and by when:

_____

_____

_____

Note: Some people are allergic to latex and some people have a fear of balloons popping! Play it safe!

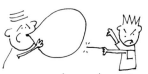

Please allow me to stress that when I talk about ideas in this book I am not necessarily talking about big ideas like inventing the airplane or launching a rocket. I am talking about ANY idea. It could be as simple as suggesting where to go for lunch. It also is not confined to the workplace. Family, neighbors, even complete strangers – they all seem to have the phrases, "well, the problem with that is" or "that won't work because" programmed to be the first words out of their mouths when they hear an idea.

Here are a couple of stories to illustrate…

Story: The Brooms
I was conducting a workshop with a Fortune 500 Company and a participant approached me after the session to share a story with me.

He told me that at his place of work there was an area right outside the men's bathroom where 10 to 12 brooms hung on a rack on the wall. Due to the place where the rack was and the type of broom, the brooms often fell off the rack and people collided with them, sometimes tripping and falling. He approached the person who had put up the broom rack, explained the situation, and offered a very simple solution. "That will take too much effort," the person who had originally put up the rack said. "Tell your guys to watch out for the brooms." The idea was to simply unscrew the rack and move it to an out of the way location! So, an idea doesn't have to be, "let's invent new brooms." It can be as simple as, "how about we move the brooms so someone doesn't get hurt."

Ideas don't have to be 'the next big thing' to have an impact – good or bad

Story: The Book

I had a friend who had written a book and was looking for feedback and ideas. One day in the hallway at work, a co-worker and I were walking past the author who asked my co-worker if he had read his book and if he had any ideas or suggestions. My co-worker said he liked the concept and had a few ideas to make it better. To this day I can't believe what I witnessed. Every time my co-worker would offer a suggestion the author would say something like, "well the problem with that is." Finally, my co-worker, in a very polite yet frustrated tone said, "you asked me for my ideas. I don't want to have a debate with you."

Imagine if the author had listened, respected, and been open to the person's ideas. Instead, it turned into a debate, the ideas stopped, and guess what – the book didn't sell. Maybe saying "that's interesting" or "help me understand" would have caused things to be different. Maybe the next idea or suggestion my co-worker was about to say was going to be the ultimate suggestion. We will never know. The fact is that although the author had asked for ideas, he wasn't open to them.

On a side note, the author I am speaking about is the same person who one time shot down my idea for a radio station that would be on the same frequency no matter where a person was in the country. For example, a person listening in Orlando listening to XYZ FM could hear the same programming as a person in Boise listening to XYZ FM. All this person told me was how it would be cost prohibitive, that I would never get the same FM frequencies, and all the other reasons my idea, 'wouldn't work.' Hmmm. Anyone hear of this new thing called satellite radio?

So what are the lessons from these two stories?
1. Respect ideas and the person giving them to you
2. Don't judge things the second they come out of a person's mouth
3. Listening works
4. Don't disrespect people when they share an idea – your reaction may show up in a book years later

Speaking of disrespect… years ago a co-worker of mine shared an idea about having our team go to New York City to see a well-known Broadway play. She had done a great deal of research on everything from the price of tickets to options for transportation. What did I do? I completely disregarded all the work she had done, criticized her idea, and pushed for something else. Her response? "Rich, you're a jerk." The sad part of this story is that for a number of years I was so caught up in the corporate rat race I did not realize the impact and consequences of my actions. To this day I still feel badly about this specific situation and others like it. She was right – I was a jerk. Who would you rather work with – a jerk or someone who is respectful and open to ideas?

This story also illustrates another point. This is my mission. This is my passion. I think about the material in this book 23 hours a day and I *still* say 'the problem with that is' or 'that won't work because.' The good news is that I am *aware* of when this happens and I am able to catch myself, instantly regroup, and refocus my energy. That is what happens to people who follow the approach in this book. I have seen it repeatedly with people who have been through my workshops. I even see it with my family. We don't have graphs, slogans, or fancy charts in our house. It has become part of the way we communicate.

Action Item #2:

Say, "that won't work" or "no, that is a bad idea" the next time someone suggests an idea. Do you see how ridiculous an action item this is? Try this and see the reaction. See if you have a productive discussion or a frustrating discussion. Ask the person how they felt.

What was the person's reaction? How did you feel?

_____

_____

_____

_____

We are all creative

The next step after creating an environment where everyone is open to ideas and respects others for their input is getting people to realize that yes, they are creative. Everyone is creative! That's right, everyone. Even you, the person reading this book – you are creative! How do I know? Because you proved it when you were a kid. We can all learn a lot from kids.

Question: Who is more creative?

An artist or an engineer?

I am so tired of hearing people say, "I'm not creative." Define creativity! Does it have to be the ability to paint, come up with nifty slogans, or play a musical instrument? I argue no. Let's think about this question. Who is more creative – an artist or an engineer? Most people would say the artist. But, when Apollo 13 was trapped in space, did NASA call in artists or engineers to save the lives of the astronauts? My point with this example is that we all possess the ability to contribute creative and innovative thinking to an organization and to society. The engineers in this example were so creative and innovative that they are featured in movies and books! [Disclaimer: I am an electrical engineer and every once in a while I feel the need to stand up for us 'geeks.'] Innovation does not have to be coming up with a fancy advertising slogan. Innovation is also needed for financial modeling and operations management. They may not be as glitzy as coming up with a logo, but they involve creativity and innovation as well – just in a different form.

Why is this so hard? It goes back to why innovation is so easy. Create a culture in which all employees are open to ideas and respect others for their input, and innovation will flourish.   Unfortunately, our culture, and by 'our' I mean American culture, is not the most open and respectful of new ideas.   And where does this come from?   It comes from school!   That's right, school. College? High School? Middle School? Let's try First Grade!   That's right. All of this fear starts when you go to school. Stand up. Be quiet. Don't do that. Do this. That doesn't go there. And the list goes on. We all can remember that kid who was a little bit different.   You know, the one that was always suggesting those 'wacky' ideas that resulted in the class laughing at him. Wait. Did I just describe first grade or my last staff meeting?

Gordon MacKenzie, in the book *Orbiting the Giant Hairball: A Corporate Fool's Guide to Surviving with Grace*, talks about visiting elementary schools.   "How many artists are there in the room? Would you please raise your hands. <u>FIRST GRADE</u>: En mass the children leapt from their seats, arms waving. Every child was an artist. <u>SECOND GRADE</u>: About half the kids raised their hands, shoulder high, no higher. The hands were still. <u>THIRD GRADE</u>: At best, 10 kids out of 30 would raise a hand, tentatively, self-consciously. By the time I reached SIXTH GRADE, no more than one or two kids raised their hands, and then ever so slightly, betraying a fear of being identified by the group as a 'closet artist.' The point is: Every school I visited was participating in the suppression of creative genius."

Story: Purple Leaves

I was conducting a workshop and one person shared a very interesting story with the group. I was telling the group about the premise behind *Mustard Doesn't Go on Corn!* when this person started nodding his head. He looked at me and very emphatically said, "you are 100 percent right." The story he told us was amazing and almost shocking.

"When I was 7 years old I was in school and the teacher asked us to draw trees," he began. "I drew some trees and started coloring the leaves purple. The teacher actually came over to me and in front of the entire class began raising her voice, telling me that leaves are not purple and I was to stop at once. She even called my parents and told them I was being disrespectful." The participants in the workshop were laughing and were a little stunned. What he said next was the punch line. "What this teacher didn't know was that we had a red maple in our front yard and to me the leaves looked purple. Luckily for me my parents were both teachers and they demanded to meet with her. They had parent/teacher meetings, my parents brought in pictures of the trees – it was crazy!" And now here is the sad part.

"To this day I am still scared to draw things or take an art class out of fear of being humiliated. And the thing is I love art."

Wow. The real kicker was that the person sharing this story was 41 years old! This event with his teacher had happened over 30 years ago and he was still upset. How many ideas has he had that he has not shared in the last 34 years? How many opportunities to 'get this off his chest' have come and gone? Forget the financials for a moment. How about the emotional aspect to this story? How about the path in life he might have chosen simply if the teacher had said, "That's interesting. Why are you coloring the leaves purple?" Instead she had to control things and in the process she placed the fear of public opinion into this person – and probably every other kid in the class. As I am writing this I am actually feeling two emotions – anger and sadness. Anger that the teacher did this, and sadness that what she did is probably happening each and every day in schools everywhere.

Wow, am I ticked off at that teacher as I write this. Deep breaths. Good air in, bad air out. Stories like this make me more focused on my mission:

> *To inspire individuals and organizations to create respectful, open, and innovative work environments that promote the creation of new ideas.*

EVERYONE – including YOU - is creative!

As a result of reading this book you may no longer tell people you are not creative. You are creative! You proved it when you figured out how to stack the chairs in your kitchen high enough to reach the cookie jar on the fridge. So, as of this moment, no more excuses.

Setting the Stage

I was watching the TV program *Inside the Actor's Studio* on the Bravo network recently. On the show, James Lipton interviews a famous actor or actress and also fields questions from a studio audience consisting of students in New York City. In this particular episode, the guest was actor/comedian Robin Williams and one of the students asked him, "how do you do it?" 'It' being Williams' ability to come up with so many ideas and be so entertaining. His response floored me. To paraphrase, "People just need to feel safe. People need to feel secure," he said. Wow. His point is simple – and humble. Yes, he is a truly unique person but, as he went on to discuss, he throws things out there knowing some will work, and some won't. Because he feels safe and secure, he can be innovative and creative. You can do 'it' too. In Robin Williams' case he is so safe and secure with himself that he can be creative *regardless* of the environment. In business, people need a safe environment.

Business people are not trained actors who, regardless of the stage, can perform on cue. The 'stage' needs to be set in order for them to 'perform' well. I remember an episode of the TV show Seinfeld where Jerry says, "The

> People just need to feel safe!

number one fear in America is public speaking. Number two is being killed. So, by that rationale, a person would rather be in the casket at a funeral than be giving the eulogy." Sharing an idea is similar to public speaking in the sense that a person has the spotlight on him when he says, "I have an idea." If the person feels safe and secure, as Robin Williams explained, idea after idea will be shared. If not, he eventually becomes, as we described earlier, 'disengaged.'

My wife always tells me she is not creative. But it is the fact that she is afraid to 'say the wrong thing' or be laughed at that keeps her innovative ideas inside. Ironically, when I make sure she feels safe to share ideas (i.e. the stage is set) she amazes me! The question is not if she is creative, but what do I need to do to make sure she feels safe, respected, and knows people are open to her ideas. *The burden is on me.*

Action
Item #3:

Get up and go look in the mirror. Based on what you have read so far, ask yourself what you are doing to promote innovation and if you are doing anything to discourage innovation. Remember, YOU are creative.

What is your response?

_____

_____

_____

Now, for you number crunching people out there, let's take this concept of people needing to feel safe one step further. Let's say we have a person, John, who had an idea shot down. For this example, assume John's total compensation is $80,000 per year. The potential costs are for illustration purposes and are based on John's compensation on an hourly basis.

Step in the process	Potential cost
John receives a negative reaction to an idea and the idea is never acted upon	Use the profit from one new product
Then…John starts to talk badly about his job and his boss to co-workers – wasted time and lower morale in 8 other people	$10,000
Then…John starts do his job less enthusiastically for the next 3 months	$10,000
Then…John starts to look for another job	$10,000
Then…John leaves (cost about 1.5 times his salary)	$120,000
Then…John talks badly about the company he left	It costs 10 times more to obtain a new customer than keep an existing customer
Total potential cost	AT LEAST $150,000

Is this worth it? All because the culture is NOT one in which all employees are open to ideas and respect others for their input. Plus, a person who is inherently creative now believes that he is not creative. Let the kid put mustard on his corn for goodness sake!

Everyone in corporations is always asking for money and resources. By NOT having a POP! Culture, companies are throwing away thousands upon thousands of dollars.

Leigh Branham, in the book *The 7 Hidden Reasons Employees Leave,* lists "Workers feel devalued and unrecognized" as reason number five why employees leave companies. In fact, Ms. Branham specifically mentions "being treated with disrespect" as one of the ways employees feel "devalued." She adds, "The desire to be recognized, praised, and considered important is our deepest craving, yet 60 percent of employees say they feel ignored or taken for granted." If you know that statistically speaking 60% of your employees or members of your team feel ignored, taken for granted, or possibly disrespected, then creating a POP! Culture where respect and openness are the foundation is imperative! Again, the impact on the bottom line is tremendous. And guess what? When employees leave, they take their ideas with them.

Question: Have you ever felt ignored or taken for granted at work?

23

Action Item #4:

In your next budget add a line item that says 'cost of not being open to ideas and respecting others for their input' and put a dollar figure next to it. Here are some items to guide you.

Item	Potential cost
Lower morale	
Lost opportunity to implement an idea that could benefit our customers	
Lost opportunity to implement an idea that could give us an edge in the marketplace	
People leave	
Lower productivity	
Total	$

This goes beyond the effects within an organization. What about when - and I can't believe I am even going to write this - a customer's idea is shot down? That would never happen, right?

Story: The doctor's office

My wife and I were meeting with my daughter's pediatrician for a routine visit. Please allow me to share this disclaimer – we love our pediatrician. He is truly an exceptional doctor and a very kind person. Unfortunately, one of his reactions, non medical related, has made it into this book. Our daughter, like most young infants and toddlers, has become frightened of the doctor's office. He suggested stopping by and getting a sticker, dropping off a note, or even just coming in for a few moments so my daughter would not think 'oh, no, here comes a shot' every time we walk through the door. Great ideas.

Suddenly another idea popped into my head. "Have you ever thought of having a cook-out or some type of family event where all the kids could come here and have fun?" I asked. "Maybe that would make your office seem like a real fun place to go." What was his reaction? Ka-boom. "Do you know how many patients we have?" he said. "That would be a logistical nightmare." Now, the shocking part is not the reaction. The shocking part is that I AM A PAYING CUSTOMER! You may be saying, 'but he is a doctor, he has lots of patients.' But does that matter? Why couldn't he simply say, "That's interesting. We like getting input from our patients." Instead, "no way" and we are off to another topic.

People even shoot down customers' ideas!

I bounced this idea off other parents – i.e. customers - who bring their child to his practice and they loved it. I told them about the reaction and they suggested, "maybe they could spread it out over a few months or do it by groups of last names." The parents wanted to make it work. Why? Have you ever brought a kid to the doctor's office? Note: Dear Dr. X – I know you are a big basketball fan so I hope you take this in good spirit and will enjoy the two free Celtics tickets you will be receiving.

Here is a question for you and your team: Do you listen to all of your customers' ideas?

Story: Creativity?

I worked at a company a few years back where I told my boss I considered myself to be fairly creative so if she knew of some projects coming down the line that needed some creative energy to please keep me in mind. The response floored me. "Creativity? Why do you need that at work?" The person then went on to tell me about the annual talent show and how that might be a good outlet for my creativity. I couldn't make this stuff up folks. It's as true as it can be. We are all creative and have the ability to be so every day. It is the culture that is the key. In case you are wondering, I didn't get involved with the talent show.

> "Creativity? Why do you need that at work?"

The Frustrating Five

So why is innovation difficult? Why do some people tell others to stop coloring leaves purple? Why do they make ridiculous statements like 'creativity, why do you need that at work?' Why do they shoot down the ideas of peers and even customers? I point to what I call The Frustrating Five™.

1. The fear of change
2. The fear of not having control
3. The fear of public opinion
4. The need to win and sound important
5. It is easier to criticize than create

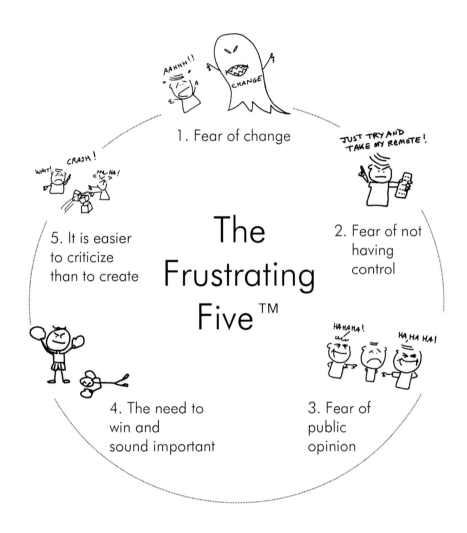

Why do I call them the Frustrating Five? I went around and around on this name and I could think of no better emotion – frustration – that describes what happens when these five elements rear their ugly head and crush innovation and creativity.

Let's take them one at a time and then discuss what we can do to get past these barriers and get on the road to innovation.

Frustrating Five # 1
Fear of Change

Has there ever been a topic that has resulted in more consultants being able to afford vacation homes, yachts, and early retirement? I thought of naming this book 'Changing the goal of change to manage changing changes.' I figured if I mentioned the word change enough I could probably crack the New York Times Best Sellers List. Keeping with the goal of this book, let's make this simple. PEOPLE FEAR CHANGE!

Fancy consulting chart and graph

Ideas =
Change

People
don't like
change!

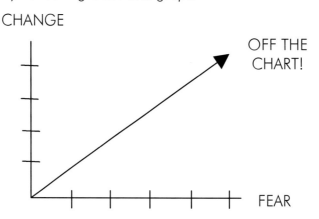

CHANGE

OFF THE
CHART!

FEAR

Presenting a new idea is presenting a change. The person hearing or reading about the new idea is thinking, 'oh no, this means a change is coming.' So how do people react? They react the same way almost all of us do when presented with change. We resist. We challenge. We try to prevent the change. This is the first reason why people react negatively to new ideas. They hear an idea, think change, and then boom, up go the walls.

So why not change! Because keeping with the status quo is so much easier. Take this action item as an example.

Action Item #5:

List out the costs you are paying for the following services:

Phone.......................... $ _____
Cable TV .... $ _____
Internet service .... $ _____
Bank fees and rates ...... $ _____
Insurance.................... $ _____
Groceries.................... $ _____

Now ask yourself if you have ever ignored the following:

- a phone call asking you to change to a lower cost provider
- a TV commercial offering great benefits and lower costs for these services
- someone telling you how to save money on these services

Here's a random thought to see how you are doing so far...
Imagine you are a TV executive and someone comes to you with the following ideas.   What is your initial reaction?

"How about this? We have two guys that run a boating tour company. The tours are simple. You know, go out for about three hours, cruise around, see the sights, and then come back. But one day the little ship sets sail and the weather starts getting rough.   Now here is where it gets exciting… except for the courage of the two tour guide owners, the tiny ship would have sunk. Instead, the ship lands on a deserted island.   But rather than make this an adventure show, we make it a comedy!   The characters are funny and over the top. We have the boat owners and a millionaire and his wife. A movie star, a professor, and another woman (not sure what she does yet). Here's where the comedy comes in… these seven folks live on this island and create their own little community.   Even though they were only going on a three-hour tour, they have clothes for months – except for the tour boat owners, we'll call them Gilligan and the Skipper, who wear the same thing every day. At the end of every show, they are just about to get saved when Gilligan does something dumb and they stay stranded. It will be a huge hit! What do you think?

The fact that *Gilligan's Island* can become a hit TV show gives hope to *all* ideas! Someone actually funded this TV show!

Frustrating Five # 2
Fear of not having control

If you ever meet someone who tells you they don't have issues with control - run. Run fast. Run very fast. I have met only a handful of people who have told me, "oh yes, I am a total control freak." Who wants to be known as the guy who is always taking over a situation when you can be known as a guy who gets things done but also goes with the flow?

Corporations are filled with control freaks. I haven't figured out if that is a good thing or a bad thing. I will admit that when it comes to managing my money, I want a control freak. When it comes to a staff meeting, I want to be around cool people.

Fancy consulting chart and graph

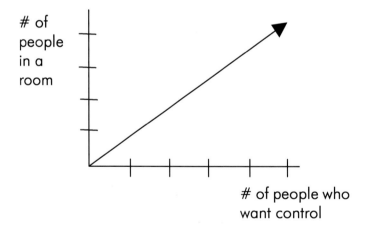

When someone hears a new idea he or she thinks 'I am not in control' right after they think 'this is a change.' So let's think about this. An idea is a change that a person can't control. Yikes. And we wonder why people shoot down suggestions! That's like hearing someone say 'let's watch something else on TV...and they have the remote! AAAAGGGHHH!

Action Item #6: This is a simple one. The next time you are watching TV and someone else has the remote ask them if you can have it. When [and if!] they give the remote to you start flipping channels. I think you get the point.

Here's a movie reference to illustrate the point. In my opinion this movie should be required viewing for anyone who works in a corporation.

Movie: "Big"

Tom Hanks plays the part of a kid who becomes an adult when he 'wishes he was big.' In the body of a 30 year-old man, he has the mind of a 13 year-old. As only Hollywood could write the script, he becomes the VP of Product Development at a leading toy company in a matter of days. In a classic scene, there is a meeting at which actor John Heard, playing the part of a mean, corporate VP, lays out a strategy for a new product. Hanks, in his 13 year-old mind, innocently offers some other options and the group begins to get energy behind some of his suggestions. John Heard, in a desperate act of trying to maintain control rather than go with the flow, begins to physically try to end the meeting and stop the discussion.

Why would he do this when the group is so energized? Why not let the idea explosion continue? Why not sit back and enjoy the fact that the group, not him alone, is ready to take action on the ideas. Why - because he is no longer in control. He wants control more than he wants the input of the group.

Frustrating Five # 3
Fear of public opinion

Let's illustrate this with an example and a question. Imagine ten people in a conference room. One is a 'crazy, creative guy who somehow slipped through the cracks and got hired.' Eight are the 'stick to the plan and don't make waves' type of people. You are person number 10. Are you going to support the 'crazy guy' or ride along with the 'stick to the plan' folks? Why make waves? Why rock the boat? Why stray from the pack? Who needs all that potential public humiliation? Most of us will side with the play it safe crew, but deep down inside we want to side with the crazy guy.

Fancy consulting chart and graph

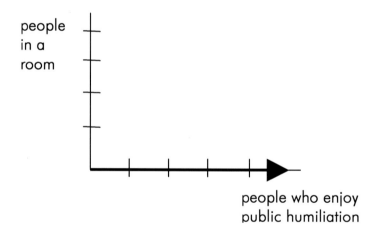

Who wants to be the 'troublemaker?' In corporations it is easier to just play it safe and glide along. First grade may be 30 or 40 years ago but those memories are forever engrained in our heads. Remember what happened to the person I wrote about with the purple leaves? **I believe people have become so scared of saying 'the wrong thing' that they simply shut down and don't say** <u>anything</u>.

Story: The Skirts

Here is a story of some very courageous kids who decided to throw the fear of public opinion out the window. One year in high school our area of the country was having a terrible heat wave. It was June, we were in an old brick building, and there was no air conditioning. I am not whining, just giving you the set up.

One day a couple of boys came in wearing shorts to try and beat the heat. In a matter of minutes, the Assistant Principal was tracking these kids down to explain to them that while there was no official dress code, the administration had made it clear that shorts were not acceptable. The kids pleaded and told the teachers and administrators that it was hot, and that by being cooler they would be more alert and do better in school. They stressed it was not a sign of disrespect, but it was a sign of frustration that the rooms were over 100 degrees. No way said the school leaders. Shorts were off limits. Now here comes the great part of the story.

> People have become so scared of saying the wrong thing they don't say ANYTHING!

The next day the kids came into school wearing...skirts! That's right, skirts! It was hysterical to see these 15 and 16 year-old young men with hairy legs and dirty sneakers walk the halls. The Assistant Principal went nuts! He made the kids go home and change. Let's think about this for a second. No rule against shorts. No rule against skirts. 100 degrees. Cooler students will probably do better that ones who are sweltering. But he sent them home.

The school was in an uproar. Could the Assistant Principal actually do this? The kids were able to bring up a new idea and generate a ton of momentum around it because they weren't afraid of public opinion. In the end the kids backed off, but who was ultimately burned by public opinion? The kids? Nope. The controlling, closed-minded, change-fearing Assistant Principal. Can you imagine this happening in this day and age? Lawyers would be brought in, CNN and FOX would be talking about it, and civil rights would be in question. All because some courageous kids had an idea, put aside their fears of public opinion, and *did something* as opposed to just complaining about the heat. Most people were thinking 'I wish we could wear shorts' but were too afraid to suggest it or do it.

Question: Do you encourage risk taking?

Action Item #7:

Try asking a co-worker to do something a little bit out of the ordinary. [It doesn't have to be 'guys wear skirts'] See what his or her reaction is. I guarantee it will be a concern of what people will think.

What was the reaction?

_____

_____

_____

Story: Laptop from the Sky

I have a friend who was asked to do a presentation on an information technology project he was leading to help sales people better prospect, target, and close business. The project was in effect a major change since everyone was being mandated to use the new system. I had an idea for the presentation called 'laptop from the sky.' I told him we could have techno music playing and lights flashing as a laptop was lowered from the hotel's ballroom ceiling. He could then deliver a super slick presentation that would blow the socks off folks. At the end I wanted to have a slide that said 'what's in it for you?' and have money drop from the ceiling with the song 'Money, Money, Money' playing. He looked at me like I was nuts. What did he do? He stood up there with the typical PowerPoint stuff and put people to sleep. Why? He was concerned about the reaction. It was funny later when the CEO of our entire business unit got

up and told us we needed to have more fun at these meetings. To this day my friend says 'we should have done laptop from the sky.'

Action Item #8:

Give a high five to the very next person you see. When you high five the person yell "Yes!" You are probably thinking 'I will look foolish' or 'the person will think I am crazy.' But let me ask you, which is more crazy – the high five or the fear that people may think you are strange because you are happy and have a great attitude?

What was the reaction?

_____

_____

_____

Movie: "Working Girl"

This movie is a great illustration of the fear of public opinion. Melanie Griffith is a smart Administrative Assistant who has dreams of being a leader in business. Her demanding boss, Sigourney Weaver, breaks her leg skiing and is out of the office for a while. In the meantime, Griffith pretends to be Weaver – all the way to the extent of suggesting new business deals to merger and acquisition whiz Harrison Ford. The whole plot blows up in her face and she goes home depressed, realizing that, in the court of public opinion, an Administrative Assistant doesn't have much clout. But they do.

Think about your company and how many 'Administrative Assistants' sit with their mouths closed because they have been led to believe they are 'just an Administrative Assistant.' What could they know, right? Wrong! The people on the front lines have more ideas than you can imagine – they just need to feel safe to let them out!

Frustrating Five # 4
The need to sound important and to 'win'

This is the one that drives me the most crazy. I wish some company out there would reward people for listening rather than speaking. This is a tough one for me, being an extrovert myself, but someone once told me 'the best conversationalist is the one who speaks the least.' Unfortunately, people have figured out a pretty neat little trick - you don't have to suggest an idea to appear smart. When someone presents an idea just start blurting out all the problems with the idea, why it won't work, and get a good old-fashioned, unproductive, volatile discussion going. Ironically, people who do this often look smart by making other people seem like they don't know their stuff.

Corporations promote the concept of innovation but the reality is that in today's world of outsourcing, downsizing, and restructuring, people have figured out that it is easier to put someone down in order to boost themselves up. It is sad but true.

Author Benjamin B. Wolman, writes in the *Dictionary of Behavioral Science* that "One major cause of workplace negativity is a phenomenon called negative attention-seeking behaviors. Negative attention-seeking behaviors are defined as socially unacceptable forms of attention-seeking behaviors, which are usually compounded with aggression. Negative attention-seeking behaviors in adults include interrupting, disobedience, quarrelsome behavior, derogation of others, uncooperativeness, demanding aid, and demanding that others accept their point-of-view." This research goes to the core of my

> The best conversationalist is the one who speaks the least.

theory on respect and openness and why The Frustrating Five are such barriers to innovation. Did you happen to catch the word 'phenomenon?'

"Winning can be losing"

These negative behaviors are often evident in individuals' need to 'win.' I propose the following to people in my workshops – 'winning can be losing and losing can be winning.' What does this mean? My point with this is that in our desire to win at games, in business, and in some bizarre way, conversations, the drive to 'win' actually causes a person to 'lose.'

Fancy consulting chart and graph

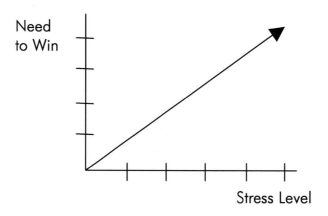

Here is an illustration. Imagine a person suggests an idea that you may not necessarily agree with. What is the natural reaction? It most likely will be 'the problem with that is,' or 'that won't work because,' or some other similar phrase. And then what happens? We are no longer in a conversation – an argument starts to take place and a 'battle' begins. And what is the typical outcome of a battle? There is a winner and there is a loser. Or maybe there are two losers. Instead of fostering an environment of respect with active listening and constructive dialogue, we enter a situation where emotions, judgment, and destructive behavior may become the norm. Why? Because each person has been conditioned to 'win.' But will there really be a winner?

Here are some potential behaviors and outcomes...
- Disrespect
- A lack of listening
- A combative culture
- Lack of trust and mutual respect
- Heated emotions
- Possible long term anger and resentment
- AN IDEA NOT EVEN BEING CONSIDERED!

This is winning?! Why the need to 'win' every time? Why can't we just let people share their ideas? Why can't we just let go of control and see what happens? So often what started out as sharing an idea becomes a confrontation.

Action Item #9:

Identify a person who is a Republican or one who is a Democrat. Go to that person and criticize something George W. Bush or Bill Clinton did, depending on your audience. Part of his or her response will inevitably be something negative about the other political party. Why? By putting the 'other side' down he or she will be able to deflect his or her party's weaknesses. Notice if the person (maybe even yourself) is trying to 'win.'

What was the result?

_____

_____

_____

Take any show on CNN, Fox News, or MSNBC that centers around people with different views talking about a topic. If you think about it, these shows are just like the Jerry Springer Show, except that instead of throwing chairs the people just throw negative reactions and inflammatory statements at each other.

Right wing host: I say we privatize social security.
Left wing host: I say we don't privatize social security.

Right wing host: It's a good idea.
Left wing host: It's a bad idea.

Right wing host: Yes we should.
Left wing host: No we shouldn't.

Right wing host: Na na na na. I'm not listening.
Left wing host: Na na na na. I'm not listening.

OK, a little exaggerated but not too far off the mark.

During the days I was writing this book a politician suggested a very radical idea for an important aspect of America's foreign policy. I will not get into the specifics of the political party or the idea. They are not important. What is important is the reaction that played out across Washington and the airwaves. I made a conscious effort to see if one – just one – member of the other political party would come forward and say, "I don't fully understand your idea. Can you explain it in more detail?" Instead, all I saw were verbal attacks, arguments, and posturing – from both sides. The idea had no chance of succeeding. I always get a kick out of the fact that most of us scoff at the Jerry Springer show and then tune in each evening to watch people with opposing views scream, yell, and argue. Why spend the money on cable TV when you can tune into the same program at work!

Frustrating Five # 5
It is easier to criticize and critique than to
construct and create

Just watch any of the latest TV 'debate' shows or a sports program. You will hear comments about what a coach should have done or what a politician needs to do differently. Everyone seems to have all the answers. Why? Because it is easier to criticize and critique than to construct and create. Have you ever noticed many of the people who criticize athletes never played sports at a high level and people who attack politicians have never run for office? Which is easier - running for political office or complaining about politics? It always strikes me as odd that we all have an opinion on politics and America in general but only a little more than half of us vote. Which takes more effort? Exactly my point.

Question: Which is easier - running for political office or complaining about politics?

Think about your 'to do list' if you have one. Do you start with the easy stuff or the difficult stuff? Most of us start with the easy stuff. The same is true when we hear ideas. We choose the easy path. Do these reactions sound familiar?
- That won't work because…
- The problem with that is…
- We've tried that before and…

These are the easy – and often ridiculous - responses. We can all do better!

Fancy consulting chart and graph

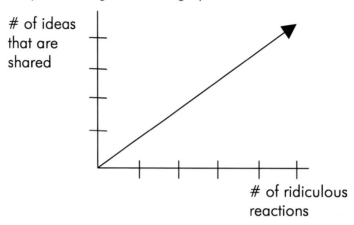

# of ideas that are shared

# of ridiculous reactions

Story: The Flight

I once had a job with a major Fortune 500 company where my role was to work on sales strategies. We were focused on ways for salespeople from different divisions to reach out to one another, leverage relationships, and start to work together on key accounts. The company sold products almost everyone could use, and by working as 'one company,' as opposed to separate businesses, everyone would benefit.

I was working with a team on the West coast and had shared with them an approach that many teams across America were using. The approach was to bring clients together and have a prominent company leader come in and talk about some of the things we were doing around productivity and leadership – topics every client would be interested in. I also told them, as instructed by my boss, to get creative, and that this was 'their group' and they needed to do what worked best for them and their clients. They came up with a super idea.

I received a phone call from a salesperson who told me the group had an idea to charter a Boeing 757, have all the sales managers invite a key client, fly to Los Angeles, serve a gourmet meal en route, and then go to see the taping of a popular TV show. It made so much sense. The plane had just about everything we sold, it was a unique idea, the clients would be 'trapped' with the other sales people for a couple of hours and the cost was about the same as having a big fancy dinner at a posh hotel. What could compete with that? Brilliant, right?

It only takes a fraction of a second to kill an idea.

I was so excited by this idea I went and told my boss. Wow, was I naive. I can't even repeat his reaction in this book. Let's just say it was something like 'golly gee, what the heck, are you crazy?' That's the edited version. Needless to say, I never even got to explain the premise behind the idea, or show the numbers the team had run to help sell it up the ladder. In a split second, the idea was dead. It was so easy to do what my boss did. It took little to no effort. But let's think about the salespeople. They put in tons of effort and tons of energy. I still wonder to this day what the impact could have been with clients.

Those sales people were seen as crazy. Now wait a minute. If a company wants true innovation so that it can ultimately exceed customer expectations and be a leader in the marketplace, don't you need some ideas like the one described here? The answer is obvious – YES! But people often take the EASY route and just tell you everything 'wrong' or 'why the idea won't work.'

I was once sharing my approach to innovation with a colleague of mine and he said, "what about the kooks? You know, the people who suggest ideas that are a waste of time." I was stunned. I rattled off a list of 'kooks' and asked him if their paraphrased ideas were a waste of time.

'Kook' 1: Michael Dell – "How about we sell computers out of my dorm room and become bigger than IBM?" (Dell Computer)

'Kook' 2: Herb Kelleher – "Let's make a bus with wings." (Southwest Airlines)

'Kook' 3: Ted Turner – "How about we show news 24 hours a day?" (CNN)

'Kook' 4: Walt Disney – "How about we build a place where we have people dress up like animals, have a parade each night with fireworks, charge a ton of money to get in, and call it the greatest place on earth?" (Disney Land)

All of these ideas were a little 'kooky' and maybe the people who suggested them were too. I am not sure since I have never met them. But look at the results their companies have produced.

And while we are on the subject of 'kooks' and wasting time, let's think about something.   How much time is wasted in companies due to:
- unproductive meetings
- gossip
- arguments because of the 'Frustrating Five'
- poor project plans
- political turf battles
- low morale

kook?

So we don't want to spend a few minutes to let a 'kook' share an idea because it might waste time, but we will spend 10 minutes around the water cooler gossiping. Interesting. And this gossip will take place in a work day where meetings start and end late, get off track, have no clear goals or objectives, consist of arguments and political battles, and even contain disrespectful comments towards others.   Who are the real 'kooks' out there? Are they the people who suggest ideas and bring enthusiasm and energy to an organization or are they the people who need to control things, speak more than listen, and put others down without understanding others' points of views?

I like to point to the prominent author and management consultant Tom Peters who uses the description 'freaks' for the people I have labeled 'kooks.' Peters notes that, "Freaks are fun. We need freaks, especially in freaky times. Freaks are the only ones that make it into the history books."

The question is not 'what about the kooks?' The real question is does your organization have enough kooks? Kooks do not waste your time. They possess the unique ability to radically improve your bottom line in a matter of seconds. Respect them and be open to their ideas. You will be rewarded.

Story: The Super Store

Here is another good example of a situation where it was so much easier to simply NOT be open to an idea. A participant in one of my workshops told me about an idea that he shared with his boss. He started the story by telling me he was 'so excited' about the idea that he had told his wife, family, and friends. His idea was to take the retail model for his business and create a 'super store' just like a Super Wal-Mart. His vision was that customers would drive for miles to experience the store environment and that it would put his company way ahead of the competition. "Yeah right," his boss said and LAUGHED at him. Smack – it was like his boss hit him with a quick right handed jab. But to his credit, the person tried again. He called a friend and shared the idea. "That will never happen," said his friend. Bam – a left hook to the gut! His energy, morale, and enthusiasm were gone. I asked him why he thought they reacted the way they did. His answer is very revealing. "I don't know. Maybe it's just human nature."

> Question: Do you have enough 'kooks' in your company?

Just human nature? Wow! It was so easy for the people to react the way they did. Imagine if they had taken a little effort to ask the person to explain the idea and - now hold on because this is going to shock you - LISTEN without judging. The idea could have been explored and who knows what the impact may have been. The person ended his discussion with me by saying, "I really don't want to be at this company any longer."

Could he have had a more detailed plan when he went to his boss or approached people differently? Sure. But it wouldn't have mattered. He never even got past the first statement.

To paraphrase a line from the movie *Jerry McGuire*, "they shot you down at hello."

Action Item #10:

Take a day to observe all of the non-productive activity you see in your company. Write each one down and log how much time is wasted. Then think if listening to a new idea for two minutes is worth it.

Here is a chart with examples to get you started. Then complete the blank version on the next page.

Non productive activity observed during one day	Time wasted
Meetings that started late, went over the scheduled time, or did not achieve the goal or desired outcome of the discussion	8 people X 1 hour = 8 hours of wasted time
Gossip and private conversations that are non-productive or even destructive	4 people X 30 minutes = 2 hours of wasted time
People working on non-work related items	8 people X 30 minutes = 4 hours of wasted time
Arguments or unhealthy debate	6 people X 30 minutes = 3 hours of wasted time
Total	17 hours of wasted time!

Blank chart for you to complete...

Non productive activity observed during one day	Time wasted
Total	

Still no time to listen to the 'kooks'?

Action Item #11:

Take note when you witness any of the Frustrating Five behaviors and observe the impact they have on the situation and on the people. Simply write down what you saw and check off which of the Frustrating Five you witnessed (there can be more than one for each situation!)

1. Example of what you saw

_____

_____

_____

2. Which off the Frustrating Five were present?
   o Fear of change
   o Fear of not having control
   o Fear of public opinion
   o Needing to win or sound important
   o Criticizing rather than constructing

3. The impact you witnessed

_____

_____

_____

Enough of these frustrating five! Let's figure out how to get past these barriers and get your company, team, or group on the simple and easy road to innovation.

Process check: How are we doing so far? Are you ready to take up the cause and eradicate The Frustrating Five from the planet? There is more to address so you don't need to answer yet.

Chapter Two

What can my team
or company do?

"There is no reason anyone would want a computer in
their home."
-- Ken Olson, president, chairman and founder
   of Digital Equipment Corp., 1977

A re you ready to have an explosion of ideas at your company? All you have to do is follow these simple steps. No fancy charts. No detailed graphs.

The steps to creating an innovative culture are simple. They are:

1. Create an Innovation Mission Statement for your company, each team, and each person

2. Spell out what will be valued and what will not be tolerated in striving to achieve your mission and vision

3. Get everyone on the same page and explain why this is critical to the organization

4. Have a process to promote innovation

5. Have fun – and don't measure it (if you can avoid doing so)

Let's take them one by one.

Step 1 of 5: Create an Innovation Mission Statement for your company, each team, and each person

Steven Covey, the famous writer and consultant to thousands of companies and individuals, stresses the idea of mission in his book *Principle Guided Leadership*. It is his view that a mission is needed at a company, department, and at a personal level. I echo his beliefs.

To become truly innovative, an entire group, team, or ideally an entire company must be passionate about fostering and promoting innovative thinking and new ideas. It can't be lip service. The company, team, or group *must have a clear mission* to make this happen. This is critical due to the power of the Frustrating Five.

Society, individuals, and companies have so many obstacles in place which *prevent* innovation that it takes a mission focused mindset to smash through these barriers. Otherwise innovation becomes just a statement at new hire orientation or the latest buzz word in the CEO's letter to shareholders.

Companies and teams have to want innovation. I mean *really* want it. That is why it has to be a mission. Think of going to the moon. It inspired people. It united people. It focused people. That is what you need to do to become innovative.

But there is a key element around this concept of having a mission that is often missing. *Each person must make it part of his or her own personal mission.* This is critical. It has to be as important as exceeding customer expectations. Why? Because people are very smart. They will pay attention and look for people who embrace the approaches in this book and also look for those who do not. Unless they see that everyone is on board, the environment will never be completely safe and open. In addition, when people make it part of their mission, they will be passionate about innovation, just as they are about customers. Innovation will not be a word – it will be something that everyone is constantly trying to promote and foster. Later in this book we will discuss identifying what is valued and what is not tolerated in terms of behaviors associated with promoting and fostering innovation. We will also review a simple process that enables your company to truly be innovative. But, as Steven Covey writes in his books, it must be at all levels, especially the personal level. This is not optional. The good news is it is simple.

Please allow me to illustrate this point. Let's take a college basketball team as an example. Why college basketball? For no other reason than the fact that I love college basketball and I had to figure out a way to work

> Innovation starts with YOU!

it into this book. Imagine the team has worked together to create the following mission statement:

"Our mission is to work together as a team, support one another, and work to our maximum potentials and abilities to ensure that after each game we can look each other in the eye and say we have done everything the best we can and have tried as hard as possible to win and represent our school in a positive manner."

Great! Who could argue with that? Now let's look at this a little deeper. Let's say the team wins a game but one person didn't hustle on a key play. Or one player yelled at a fan. Or one player was late for the game. The team won but are they achieving their mission? The answer is no. This is why the mission to become innovative must start at an individual level. 'But the team won,' you might say. 'No big deal.' Let's say the team loses. Still no big deal? This example can be translated to the workplace. You might have a person who says, "Who cares if I refuse to change, refuse to give up control, and I am disrespectful to others when they present ideas. We made our numbers." This is the same as the team that won the game but really didn't achieve the mission. If your mission statement to achieve innovation is only at the company level (or if you don't have one at all!) and it's not specific to each team and to each person, the odds of overcoming the Frustrating Five are low.

You must have an innovation mission statement for your company, your team and for each person. It is not the same as your company's, team's, or each individual's overall mission statement. Too many companies make the mistake of making a blanket statement like 'we will be innovative,' tape it to the wall (in a very nice frame to make it look very important), and expect the Frustrating Five to disappear. The Frustrating Five are much too big for one statement in an annual report or on a company poster.

Note: If you are leading a team or department, you may want to start with your own folks. Changing the entire company is a big fish to fry.

Here are examples of Innovation Mission Statements for a company, a team, and an individual.

Company's innovation mission statement
Our innovation mission is to create a culture and work environment in which all employees encourage, value, and respect the ideas and suggestions of other employees, clients, and business partners in order to constantly exceed customer expectations and improve our bottom line.

Team's innovation mission statement
Our innovation mission is to practice and promote innovative behaviors and processes within and outside of our team environment in order to dramatically increase the confidence and self esteem of our colleagues by

> The Frustrating Five don't go away without a fight.

showing a sincere willingness to listen to, respect, and ultimately act upon others' ideas. Our mission will be to excel at our business objectives while respecting and encouraging the ideas of others.

Individual's innovation mission statement
My innovation mission is to continuously be open to ideas, model the key behaviors our team has agreed upon, and always *defend and support those* who fall victim to any of the Frustrating Five.

See? It's simple! And notice that the focus on business results does not disappear – it is part of the mission. This is a very critical point. There is a very basic cause and effect relationship as I mentioned earlier in the book. The goal is to create the right culture for innovation. By doing so the result will be a tremendous return on your investment. **The goal is to increase your bottom line.**

Action Item #12: 	Write down the dates, names, and other information needed to get the Innovation Mission Statement process started… NOW! Use the worksheet on the next page.

Mission statement	Will be written by when?	Who needs to be involved?	What help will be needed?
Company (if applicable)			
Team, Department, or Group			
My own			

Use this space to write a draft of your Innovation Mission Statement.

_____

_____

_____

_____

_____

You may be asking, "Why are we writing this down? Can't we just say 'OK, I get it.'" No! Have you ever tried to go on a diet? Start an exercise program? Achieve a personal goal? Step one is to write it down. It becomes visible and it becomes real.

Super! You are on your way to creating an innovative culture. Remember, you are creative and have the ability to be innovative. Now it is just a matter of getting the ball rolling with the whole company, team or group.

Step 2 of 5. Spell out what will be valued and what will not be tolerated in striving to achieve your mission.

I can't stress enough how important this step is to the process. Not just writing down the words, but living it each and every day. I also can't stress enough that this is not a 'police action' or a way to intimidate people. It is a way to know exactly what is expected and what will not be tolerated. Why is this so important?

Every company has mission, vision, and values statements (or they should). Just as I suggested in the previous section, there needs to be a set of values and a definition of what will and won't be tolerated in terms of promoting innovation. For example, many companies will have values that are very broad and high level.

One value that companies have may be something like: "We will be a steward in our community."

Who can argue with that? But what if that statement was: "We will be a steward in our community by allowing our employees to take 5 days per year to help with civic and socially responsible organizations."

Hmmm. That gets my attention. Now what if we added, "These five days will be incorporated into each person's performance review and will be as important as any other performance measurement." That's interesting.

Now let's take it even a step further. "We will not penalize employees in any way for participating in civic and socially responsible endeavors and we will not tolerate any associate or manager who discourages or creates obstacles for individuals who are completing their community service. We, as a team and a company, will work together to ensure the needs of our clients are met while also being a steward and role model for our community."

Wow! We have gone from something at 30,000 feet to ground level. I ask you, which of the statements above will have a bigger impact? My bet is on the latter. What do you think?

Earlier I wrote about how companies need to really *want* innovation. Taking the time to clearly spell out what will be valued and what will not be tolerated is a small fee for the tremendous ROI companies will see by making sure everyone knows the rules of the game. Hint: Get everyone involved in making up the rules – they will enjoy the game much more than if the rules are imposed on them.

I also mentioned company politics and turf battles. I *know* these don't happen where you work, but play along. Listing out specific values enables people who need to feel safe to know that they are in a culture in which they have the freedom to share ideas and NOT be shot down the second the idea comes out of their mouths. Also, it forces people who have a tendency to dominate meetings and try and control the outcome of conversations to take a step back – and listen! These values are sort of a 'Bill of Rights' for people (cue up corny patriotic music here).

Here is an example of an Innovation Bill of Rights. Feel free to adapt it for your company or better yet, make your own!

Innovation Bill of Rights

One of our core values is to be open to new ideas and respect others for their input. Every employee has the right to expect that as a company:

- o *We will provide people with an environment that allows each person to share ideas without concerns of any kind.*
- o *We will listen to ideas and not pass judgment without fully understanding them.*
- o *We will treat ideas as gifts that should be embraced and valued.*

Individuals who help foster ideas and encourage others to share new points of view will be considered one of our most valuable assets - and will be recognized as such. People who do not exhibit behaviors of respect and openness will be given the opportunity to be coached and provided needed tools and resources to help modify their behaviors. We will not tolerate behaviors that stifle creativity or innovation.

**Action
Item #13:**   Write down what you will value in terms of innovation and promoting a POP! Culture:

_____

_____

_____

_____

And what you will not tolerate:

_____

_____

_____

_____

Hopefully I have shown how this Bill of Rights is much more powerful than one line in a company's values statement that says 'we will strive to always be innovative.' The true test however will be recognizing those who follow the values and working with those who do not.

Story: Billy

I was working with a leading non-profit organization for which I was conducting a workshop to kick-off a 3 day strategic planning off-site with their senior staff. I was told by my primary contact and sponsor of the session to be aware of a person named Billy. I was told he was a very negative person and that he would most likely be very cynical during the workshop. We finished the workshop and I left the folks to continue their three days.

I followed up with my contact and asked her how things went and if the group applied the tools and behaviors they had learned and developed. "It was amazing," she said [and I am paraphrasing]. "There was a point in the meeting when we were bouncing around an idea and Billy said, 'I know I can often be seen as the 'well the problem with that is' guy but can I share some concerns?' The group was stunned, smiled, and laughed with him. He was *asking* if he could share concerns. It was a great show of respect for others. He was smiling too, which made everyone relax. As a result of what he learned in the workshop, Billy was able to diffuse the situation and take a step back to make sure he did not shoot down the idea or be disrespectful to the others in the room. What normally would have been a tense conflict turned into a healthy and productive discussion."

There is an even greater punch line to this story. I later found out that Billy was the person in charge of budgets and finances for the group so his filter was usually one of 'what is that going to cost.' The tools and insights gained from the workshop were able to get him to a point of not thinking about the dollars right off the bat when an idea came out, but to listen, digest the information, and then have a constructive response. Perfect! How much time did my workshop take? About 2 hours. What was the return on investment? A productive 3 days! The Frustrating Five had been defeated – for now. The pursuit of the mission must continue because the Frustrating Five are resilient – they do not give up easily.

An Innovation Bill of Rights or even a simple list of what will be encouraged and what will not be tolerated will pay great dividends for your team or company. It gives everyone a voice and sets clear expectations for everyone involved.

What about the people who won't follow the Bill of Rights? This is a tough question and one you will have to face head on. I have my own opinions on the subject but you need to do what works for your team. Hint: Have the team work with you on this action item.

Question: What about people who are good performers but shoot down ideas?

Action Item #14:

Answer this question – what if I have a person who has a great performance record and strong abilities but does not embrace our values?

_____

_____

_____

_____

_____

_____

_____

_____

Step 3 of 5. Get everyone on the same page and explain why this is critical to the organization.

Now that you have clarity on your mission and values, it is time to make sure everyone is on the same page. 'Everyone' in this case could be your team or hopefully your entire company. I tell clients that my approach to innovation is not like signing up for a class or going to an 'open enrollment' training session. EVERYONE has to participate or it is a waste of time. There is no use in sending one person to an innovation off-site only to come back to work and find the Frustrating Five still running rampant. The Frustrating Five are too big for one or two people to defeat. It has to be a team effort to create a POP! Culture. Again, they key is the culture.

While working at GE I saw this done brilliantly with various initiatives. Tools and approaches like process mapping and change management weren't courses in a catalog. They were long-term initiatives that were designed to promote everyday practices that would become second nature. They were not optional and everyone went through a process for learning how to apply various tools *everywhere* in the organization. I compare this to everyone speaking a common language. How much easier it is to get things done and to communicate when everyone is on the same page? Plus, it prevents artificial turf battles from arising. Statements like 'only managers need to go through this workshop' are non-existent when the approach in this book is followed.

The other thing GE excelled at was creating the 'wake up call' for people to know why these changes and initiatives were important and how they would impact business results. While they did say 'we are doing this,' they also explained why and related that to the bottom line. Without a sense of urgency or the answer to the question 'why are we doing this,' people will resist change and will think 'this whole innovation thing is just a fad.'

Imagine sharing this statement with your team or company:

> "We are taking on this innovation mission due to the fact that our competition has dramatically raised the bar in terms of their ability to provide world class products and services to our customers. Due to the efforts of our competition we are losing market share at a rapid pace. If we are able to tap into the creative energy of EVERY employee and develop an environment where everyone is open to ideas and respects others for their input, we will have an 'idea explosion' and be able to implement new and radical ideas that enable us to remain the leader in our industry."

I was once at a meeting where the CEO of a major financial institution was speaking about a new technology initiative. "If we don't do this we will be out of business in two years," he said to the employees. The room was silent. "But if we do this we will be successful beyond our wildest imaginations."

Action Item #15: Answer the question... why are we creating a POP! Culture:

_____

_____

_____

_____

_____

And... what if we don't?

_____

_____

_____

_____

_____

Your answers should be actionable and create a wake-up call for the team! Think bottom line and ROI.

In addition to the questions 'why are we doing this' and 'what if we don't' there are three other questions that need to be answered:

  A. What's in it for me?
  B. How will this affect me on a daily basis?
  C. What will this look like?

Let's take these one at a time.

A. 'What's in it for me?'

Is there a bigger question we all ask ourselves when we hear an idea, a proposed change, or something new? Sure, we would all like to say, 'oh, I never think that way' but we do. We all do. When an idea is proposed or a new concept or approach is introduced, people inevitably get to the point where they ask, 'What's in it for me?' It is human nature.

_____

People always think 'What's in it for me?'

_____

Let me put it this way. In order for me to get you to join me on my mission I have to think what's in it for you, the reader. Otherwise it is about me, me, me when it has to be about you, you, you.

As I mentioned before, it is the split second an idea is shared that we are primarily concerned with. With that said, if you can beat the Frustrating Five in advance then by all means do so! Remember that word 'respect' we talked about? It goes both ways. You have to respect the people who are *getting* the ideas as well. The way to do that is to always think about things from the other person's point of view and ask, 'what's in it for me' with me being them, not you, or is the pronoun I, or we...you get the point. People want to know, 'what's in it for me?'

Action Item #16: What is in it for your team or company if you create a POP! Culture?

_____

_____

_____

B. 'How will this affect me on a daily basis?'

People will say, 'Okay, I get it. This is good for the company and good for me. But what will my day-to-day job look like if we have a POP! Culture?' This question is a derivative of 'what's in it for me?' People need to know what is going to change (for the better!) and how their jobs and their lives will be affected. Will it mean longer meetings? Will there be reports? Is this part of my performance review? Will my role in meetings be changing? These are all potential questions that will be racing through people's heads when they hear about a POP! Culture. Beat the Frustrating Five in advance by answering these questions.

Action Item #17:

List out the questions and possible answers so people can know how a POP! Culture will affect their day-to-day jobs.

_____

_____

_____

_____

_____

C. 'What will a POP! Culture look like?'

People are visual. They need a picture. This ties into your Bill of Rights and could even be called a *vision* of what your team or company will look like as a result of implementing a POP! Culture. If you want to use crayons and draw smiley faces feel free!

Action Item #18:

Write out a 'picture' of what your team or company will look like once you have implemented a POP! Culture. Use some of the information you have read or action items you have completed.

_____

_____

_____

_____

In summary, here are the key questions that need to be answered to ensure everyone is on the same page:

- o Why are we creating a POP! Culture?
- o What if we don't?
- o What's in it for me?
- o How will my day to day job be affected?
- o What will this look like?

In addition to working on the questions above, what I do with clients is to put everyone through a fast, high energy, and fun workshop that creates a 'common language' that everyone (and I stress everyone) is speaking every day on the job. The 'language' they are speaking will be explained in more detail in the next section. In my workshops everyone gets a beanie cap. Everyone moos like a cow. More on that later.

Step 4 of 5. Have a *process* to promote innovation

I have found in the course of my experience that one of the ways to get people to be more open to change and receptive to new ideas is to provide a roadmap or process they can follow. With that in mind, I have developed a process for innovation that I call NEWIDEA!™ The goal of NEWIDEA! is to help teams, groups, and companies have a clear path of the steps they can take and the behaviors they can model to enable a POP! Culture to become present in their organization.

The NEWIDEA! process is outlined on the following page. The idea here is that if people follow these seven simple steps, innovation will flourish. It is that easy.

NEWIDEA! Process

N – No negativity!

E – Encourage the person

W – Wait…and LISTEN!

I – Include input

D – Document the idea

E – Evaluate and explore options

A – Action!

Here are some common questions I hear regarding NEWIDEA!:

Question: "Does every idea go through the NEWIDEA process?"	The answer is no. What I tell people is that if you are able to get your folks to model N, E, and W, you are 90% of the way to achieving great things. I describe this process as being 'situational,' meaning that the

situation will determine how far along the process you will travel.

For example, let's say a friend of mine and I are on a subway talking about our jobs and I say I have an idea. I don't expect him to say, "hold on, we better get off at the next stop and get a flipchart and some markers so we can write the idea down." That would be crazy. At the same time, if my friend and I are meeting with the Board of Directors and we are trying to come up with ideas for the strategic direction for the next three years, a flipchart and some markers may not be a bad idea. Low structure – make sure to get through N-E-W. High structure, pull out the flipchart and go all the way to A - Action.

There is one part of NEWIDEA! that needs particular attention. W – Wait and Listen! If I was planning to add a sixth item to the Frustrating Five it would be that people don't listen. Oh sure they may nod and even say an occasional 'hmm' but are they really listening? Most of us are doing what? We are preparing our response! That is not listening. That is hearing. Listening is respectful. It shows openness. It is at the core of innovation and creativity in organizations. In my workshops I spend a good deal of time on fun exercises that get people to listen and understand what is being shared.

Here's a thought: A person who listens to an idea is as valuable as the person who shares the idea.

Here are some statistics from www.listen.org:

- Amount of time we spend listening? 45%
- How much we remember of what we hear? 20%
- Amount of us who have had formal educational experience with listening? less than 2%
- We listen at 125-250 words per minute, but think at 1000-3000 words per minute

These statistics underscore the importance of listening with respect to creating an innovative culture. The statistic I find most revealing is the fact that people listen at 125-250 words per minute but think at 1000-3000 words per minute. That, combined with the fact that less than 2% of people had formal education around listening, is a strong argument that companies need to spend more time on teaching people listening skills than presentation skills. Listening to an idea is as important as the original suggestion itself.

Action Item #19:

Imagine I am sitting next to you as you are reading this book. I say, "excuse me, can I suggest an idea? I would like to suggest that we give all the power in this country to the Republican Party. Write down your first thought.

_____

_____

_____

What did you write down? My guess is about 51% of the population will respond with 'that's great' and 49% will respond with 'you're nuts!' But how many will respond with, 'Why do you suggest that? I want to make sure I understand your idea.' That is listening. That is respect. That is being open. That leads to innovation.

I could switch Republican Party to Democratic Party and the answers would be the same. The point is that so often we hear an idea and are so quick to judge or prepare a defensive response that we fail to take the time to listen and understand the idea. And then we wonder why arguments, finger pointing, and out right dismissal of ideas is so prevalent. Take a moment to listen.

Story: The Styrofoam™

Here's an example of where this process and a little bit of good old-fashioned listening could have helped. I am part of a non-profit organization whose goal is to help homeless children. Yes, homeless and children are in the same sentence. We were having a meeting about an upcoming event and a person on the team shared an idea with the group. She mentioned that her husband was able to get large sheets of Styrofoam for an upcoming event. Her idea was that we could use the Styrofoam to create murals and signs to really dress up the event space since the location is typically in a very drab building. You would have thought she suggested shutting down the organization! "They will be tough to transport." "Painting them will be difficult." "Styrofoam doesn't look good." This was just the start of the flood of reasons why her idea 'wouldn't work.'

Imagine if the group, recognizing we were in a fairly informal setting, had just followed N – no negativity, E- encouraged her to continue, and W – waited and listened. That's all it would have taken. Why? Because it turns out the Styrofoam was FREE and she already had answers to most of their questions. Instead, by the time those key facts came out the group had already explained all the reasons why 'the idea wouldn't work!'

Question: "Does every idea get implemented?" The answer is NO! The goal is not to say yes to every idea or to build on every idea. It is simply to be open to ideas and to respect others for their input. That's what a POP! Culture is. Openness and respect. Openness and respect. That's what this process does, especially in the first three steps. By having a POP! Culture, people will not expect every idea to be acted upon. The combination of a POP! Culture and the NEWIDEA! process encourages people to keep 'throwing out the ideas' without concern of becoming victim to the Frustrating Five. Remember, the more ideas, the higher the rate of success, and the better the chance of *improving your bottom line*.

Question: "How do we model each step of the process?" That is the beauty of this approach. *Your* team, group, or company brainstorms the behaviors, tools, and techniques that work for *you*. There is not a pre-packaged list of tools and behaviors that are given to people. Why take this approach? Simple. Who knows your culture best? You and your employees. Do people like a) being told what to do or b) being part of the process around a change? Note: in case you missed it earlier – people fear change! I will go with option b. Plus, it allows you to begin using the NEWIDEA! process to jumpstart the initiative.

**Action
Item #20:** Conduct a meeting with your team to brainstorm ways to model the behaviors in the NEWIDEA! process.

N   Ways we will ensure people will react without NEGATIVITY to new ideas.

"The problem is...

_____

_____

_____

_____

E   Ways or things we will say to ENCOURAGE people to continue with their idea.

Wow!

_____

_____

_____

_____

W   Ways we will promote WAITING and listening within our team or company.

_____

_____

_____

_____

I   Ways we will promote building on ideas and adding constructive INPUT to suggestions.

_____

_____

_____

_____

D   Ways we can help ensure ideas are DOCUMENTED and visual.

_____

_____

_____

_____

E   Ways we can explore options and EVALUATE ideas (not in the split second they are shared!)

_____

_____

_____

_____

Pro's | Con's

A   Ways we can take ACTION on ideas.

_____

_____

_____

Remember, change takes time. With that in mind, in my workshops I use things as crazy as squeaky hammers to 'smash' negative reactions and I encourage people to bring the hammers to meetings and to use them when someone shoots down an idea.

I also use many visuals, some that are tangible (beanie caps and participant guides) and some that are actions that are simple to model (high fives and mooing like a cow). The goal is to do almost ANYTHING that will enable the approach I have outlined to become embedded in a company's or team's culture and to make the process and tools simple and easy to use.

"Squeaky hammers? Mooing like a cow? That sounds unprofessional," you may be saying. No, laughing at someone's idea, judging it without understanding, being disrespectful – that sounds unprofessional. Why do we tolerate these negative behaviors but display caution for something like a squeaky hammer or mooing like a cow? Why do companies allow negative behaviors to take place every day but are fearful of 'what people will think' if they pass out beanie caps? This is at the core of what a POP! Culture addresses, and what needs to be confronted head on to be a truly innovative company.

Question: "Why write the idea down? Can't we just talk through the idea?"

People like to be recognized in a positive light. People are also visual. By simply writing down an idea you accomplish a number of things:

a. You make sure what you write down is actually the idea the person is trying to express.
b. It validates the concept that the group is open to ideas and respects others for their input.
c. It gets people to focus on the idea, not the person who said it. This is HUGE. By writing an idea down, the focus of the entire group moves from the person who shared the idea to a piece of paper/white board that everyone can see. Imagine if there are two people in a meeting who have a history of arguing with each other about EVERYTHING. Writing the idea down takes the focus away from the people and allows the group to put its energy into focusing on the idea, not a potential altercation.

Here's an example of what can happen when an idea is not written down.

Story: New Hire Orientation
I was at a meeting when I worked for a large financial services company and the question on the table was "how can we dramatically improve our new hire orientation process?"

Part way through the meeting I said, "I have an idea" and I suggested that all new employees spend at least one day in the mailroom. I actually called it 'Day in the Mailroom' for all new hires. Everyone in the room pretty much laughed or scoffed or said that wouldn't happen – the typical responses – and the facilitator didn't write down my idea. I said, "No, please write down my idea. I really think this is something that would be good. Could you please write down Day in the Mailroom?" He just ignored me, brushed me aside and said, "Let's keep going with the meeting."

He would not write down the idea. I was passionate about the idea so I asked again. "Fine," he said and wrote 'Day in the Mailroom' on the flipchart. I said, "Thank you. Now, if I could explain why I suggested this idea, the group may find it interesting. We are a very large company that is actually many, many small companies all under one umbrella and we all happen to be in the same location," I told the group. "If you notice when you go around the building and come off the elevators, the first things you see are the mailboxes for all the staff throughout the company. My thought is that the mail people know where people sit, who the people are, and what the various people in the company do more than anyone. By spending a day traveling around the building, a person would get a sense of the company's bigger picture, would meet a whole lot of people, and have fun along the way. It would be a fun, inexpensive and highly visible way to get people to know about the company and increase the visibility of new hire orientation."

The room was stunned and suddenly people started to say things like, "That sounds interesting" and "that isn't such a crazy idea." Interestingly, the facilitator ignored the energy and moved on.   The good news is that by writing it down it did raise the energy in the room.   In addition, it enabled the focus to be on the idea, not on a potential argument between the facilitator and me.

Why not just write the idea down? Why pass judgment? Why the need for control? The result of writing an idea down can be exponentially higher than not writing it down. In fact, I am struggling to find one good benefit except for possibly saving money on markers and flipchart paper to not writing down an idea.

In reality this facilitator wasn't open to new ideas, even though – and this is the huge lesson – the meeting was advertised as "please come to this meeting to share creative ideas - all ideas are valid." Don't write ideas down because you have to.   Do it because it makes sense.

Action Item #21: Get a flipchart and markers for every conference room and office – and use them!

I will do this by when: _____

Question: "Won't the NEWIDEA! Process take a lot of time?" The answer is NO! This is about identifying, modeling and integrating key behaviors into your culture. It will just become a natural way for people to conduct themselves. Don't get bogged down in the process – focus on what works for your team or company.

For example, I was working with one of the largest restaurant franchise companies in America at a time when they were making a huge investment with their management team on various principles of leadership and overall life management. When planning for what we would do to create a POP! Culture and introduce the NEWIDEA! process, we made sure that what we did *complemented* what was already taking place. For example, a big part of their management effort was around affirmations and visioning. Great. We simply built that into the process. We focused on what would work for their company, their culture, and what would be easiest for their employees. It was a great success.

Question: "What about the money we have invested in learning brainstorming and other idea generation tools?" The good news is the time and money you may have spent on these various innovation and creativity initiatives is not lost. In fact, if you have spent the time, energy, and money to learn various approaches like brainstorming – GREAT! What I am proposing is that brainstorming tools and techniques are effective ONLY if you have a culture and environment in which everyone is open to ideas and respects others for their input AND you have a process to take the ideas through to completion. Without the culture and environment in place, people can actually get frustrated and feel that the time spent on learning how to generate ideas was a waste. When you start to hear people say things like 'why did they ask us for our input and teach us brainstorming if they aren't going to use our ideas,' that is a good clue that there is still work to be done. As I mentioned before, do what works for your team and organization. If you already have some nifty processes and techniques in place, super! I teach three techniques in my workshops, but explain that they can turn out to be useless without a POP! Culture and the NEWIDEA! process.

Question: "Does this mean that we never have a healthy debate around ideas?" NO! Healthy (and I must stress the word healthy) debate is still welcomed and NEEDED. What I am suggesting is that debate and potential arguments don't need to be the first thing out of people's mouths when they hear an idea. There is plenty of time for healthy and respectful debate and discussion in the second 'E' portion of the NEWIDEA! process – Evaluate the idea and explore options.

Please allow me to illustrate this concept. Let's say there is a meeting and the question on the table is 'how can we increase sales?' A person suggests, "let's go visit all of our customers." People in the group may respond with reasons why the idea 'won't work.' They also may respond with questions like "how much will that cost?" This brings me to a very, very important point. I have never met someone who can complete a cost benefit analysis, provide a comprehensive business plan, and have a tangible list of action items – IN ONE SECOND. If a person shares an idea in a brainstorming meeting, guess what? They probably don't have all that information! So, rather than come at them like the District Attorney on an episode of *Law & Order*, maybe it would be better to wait until everyone UNDERSTANDS the idea and has let the person finish before quizzing them on how much nine plane tickets to six clients in a four day time period will cost – building in compound interest and inflation. Okay, forget the compound interest and inflation part, but you get the point.

Healthy debate and discussion – yes. The split second an idea is shared – no. As I mentioned before, the primary focus of this book is on the *split second* an idea is shared. It is great if the person sharing the idea has a chance to do some homework ahead of time. Most times this is not the case, especially in brainstorming sessions or meetings in which people are asked to work on a problem.

I have friend who asked me to help him come up with some creative ideas for an invention he had conceptualized. I shared a suggestion and he said, "That will cost too much." I asked him, "how much?" He said, "I don't know, but is sounds expensive." He had no idea! Why not let me finish the idea, try building on it, and then ask, "Do you think that would effect the cost?" Instead we are about to have a debate and potentially an argument at the absolute worst time – one second into the idea.

Story: The Circus

Here is another example of how the NEWIDEA! process can work. I was conducting a workshop with a leading global financial services and insurance company. The group I was working with was having a tough time with the perception that their team was a roadblock to getting things done. This was not reality at all. During the workshop we were focusing on the following real life situation: "How can we create a valid perception within the field and other departments that we are a necessary and valuable step in the process of serving our customers?"

I told the group they needed to come up with at least two ideas that were 'ridiculous' and 'outrageous.' I do this with groups to show that what is often perceived as ridiculous is often not ridiculous at all. One of the people in the group had written down on a flipchart 'have a circus at our department.' I asked, "is that your ridiculous idea?" She said, "yes" as everyone laughed. I said, "okay, let me show you how using the NEWIDEA! process will take what you think is an outrageous idea and actually show you it is not so outrageous at all."

I said, "Please say your idea again." She said, "let's have a circus at our department." "Great," I said, "This is where we start with N – No negativity." I looked at the group. No laughter. No mocking. No phrases like 'the problem with that is.' "Let's move on to E – Encourage the person. Why do you say let's have a circus?" I asked, also pointing out that I was going to model W – Wait and listen. What she said was outstanding. "A circus is loud, fun and a little crazy. My thought is that if we had a circus going on people would come by and see what all the commotion was about. Then, while they were here, we could show them what we do and explain to them face to face how we fit into the process. Plus, it would be fun and different and they would have a good time and tell their co-workers to go to the circus."

Suddenly 'let's have a circus' didn't seem like such a ridiculous idea. We had gone through N, E, and W and by doing so we had learned something. By being open to an idea, respecting the person for her input, and having a process to follow that promotes innovation and creative thinking, we were on to something. You could see and feel the energy in the room.

"Now let's move on to I – Include input," I said. "How about if I build upon your idea of a circus? What if we have some type of contest or prize for people?" Now people were getting even more excited. "Let's write the idea down," I suggested, moving into D – Document the idea. I wrote down 'let's have a circus' and then to emphasize the power of the NEWIDEA! Process, I crossed it out – with her permission of course. I then wrote 'let's have an event that creates a buzz in the field and in the home office that inspires people to come visit our department so we can then talk with them face to face and show them how we add value.' I asked, "Is this your idea?" "Exactly," she said. "So it's really not about a circus, is it?" I asked. The participants were shaking their heads. I also asked, "where is everyone focused right now?" We were all focused on the flipchart – not on the person who had suggested the idea. Why? Because the idea had been written down.

Another lesson from this story is that 'let's have a circus' was no longer her idea, it was the group's idea. I call this 'Mi idea es su idea'™ or 'My idea is your idea', where an idea no longer is associated with a person, but has the ownership of the group. I will get back to that in a moment. Continuing with the NEWIDEA ! process, we went into E - Explore options and evaluate. I asked, "What tools do you use to help you make group decisions?" My goal here was to use what was already part of their culture. We worked through the idea for a short while and then I said, "Okay, now let's move to A – Action. What should we do from here?" We decided that we would have three people take the idea and all the work we had done, flesh it out further, and come up with a plan within 30 days to present back to the group. Everyone was excited. Note: If by exploring and evaluating options the group decided to take NO action, that would have been fine as well.

'Mi idea es su idea' – My idea is your idea.

We weren't having a circus (although I personally think that would be hysterical), we were 'creating a buzz in the field and in the home office that will inspire people to come visit our department so we can then talk with them face to face and show them how we add value.' It wasn't about a circus at all. Imagine if the typical reactions like 'we could never bring an elephant into the building' or 'that would be messy' had come out of people's mouths the second the idea was suggested. The focus would have been on a circus, not on what the person was really trying to say. Everything we had just done was simple and easily transferable to ANY meeting. In a matter of minutes the participants saw the effects of a POP! Culture and the NEWIDEA! process. Everyone now had a common language to share ideas and a simple process to get them to action. Fast, easy, and energizing. The boss was a hero.

The NEWIDEA! process is critical because when someone suggests an idea, it is just that – an idea. The person most likely does not have all the details. The person has usually not had time to completely think it through from A to Z. And often, it is not exactly what the person is really thinking. So, with this in mind, why, when an idea is suggested, do people respond with questions or discouraging comments? What is the harm in simply being open to the idea and respecting the person for their input? None! There will be plenty of time for evaluating and exploring options. This is a critical point. The approach I am suggesting does not mean that healthy debate goes away. What I am proposing is that debate and judgment not be the very first reactions to suggestions or new ideas. My approach is NOT saying 'yes' or 'yes and' to everything – that would be crazy. With a POP! Culture people know that not every idea will be implemented, but that they are in a safe environment to keep tossing out suggestions. Then NEWIDEA! enables them to have a roadmap for ideas to follow. Everyone needs to chill out, take a step back, and let people explain their ideas! Pause. Take a deep breath. Okay, I'm calm now. Let's continue.

Let's get back to this concept of 'Mi idea es su idea' or my idea is your idea. We are all so caught up in 'who will get the credit' or 'who will get the recognition' that we often become our own worst enemies. I compare this to an athlete who is more concerned with personal statistics than winning the game. What good is it to be an all-star on a losing team?

Also, I am firm believer that if you want to get an idea through the system, you have to make people part of the process, give up control, and let others be involved. It has to be their idea. This is part of the reason that people react the way they do when presented with new ideas. It goes back to 'what's in it for me' and 'how will this affect my day-to-day job.' I can prove this with my daughter and her bedtime.

My wife and I try to stick to our guns when it comes to putting her to bed. Unless there's something wrong, we try to let her cry it out if she's having trouble getting to sleep. We thought we were doing a good thing by closing her door so noises like the phone ringing wouldn't bother her, but she would sometimes get very upset. Then one day my wife and I asked her, 'do you want to close the door?' Well, guess how much easier bedtime has become! It's neither my idea nor my wife's idea to close the door, it has become *our daughter's* idea to close the door. She is part of the process. She has input. She has an element of control. And, she always closes the door! It's amazing what can happen when the 'Mi idea es su idea' approach is used. And forget all this innovation stuff – we get more sleep! I say this jokingly but it is a very important point. The goal is to get my daughter to go to bed without a fuss. Plus, an added benefit is we get more rest. So, why would I even care if closing the door was my idea or not? Sounds simple, right? Then why - and we all do this - don't we let others have 'the glory' or 'the credit' if ultimately the result is what you need or want to happen?

I often tell people in workshops, particularly managers who are looking to implement a change, that the key is to have the idea come out of their employees' mouths, not the managers' mouths. Buy-in and involvement will increase, resistance will decrease, and more ideas will be implemented.

Here is a great quote that someone once shared with me:

> Go with the people.
> Live with them.
> Learn from them.
> Love them.
> Start with what they know.
> Build with what they have.
> And of the best leaders,
> when the job is done, the task accomplished,
> the people will all say: "We have done this
> ourselves."
>
> *Lao Tsu, China, 700 b.c.*

The next time you have an idea or need to implement a change, try and make it the idea of others, not your idea. Is it credit and recognition you want for your ideas, or for your ideas to be implemented? Maybe you can get both, but is credit really the goal? Think "Mi idea es su idea" and you will be amazed at how barriers come down and how many ideas move forward. How about adding this to your Bill of Rights?

Imagine a culture where people know and accept that once an idea is out there or on a flip chart it is no longer their idea but rather the 'ownership' of the group. I put ownership in quotes because I don't want people to think copyrights. Ownership in this case is meant to imply that the *team* owns the idea, not the individual.   Plus, I would argue that the satisfaction from giving of yourself and seeing the team succeed will be much greater than any recognition you get – alone.

Okay, that is pretty heavy stuff. Let's get back to a lighter topic!

Step 5 of 5. Have fun – and don't measure it! (if you can avoid doing so).

This is the step in the POP! Culture approach that is a lightening rod. It brings up two items that often go against the grain of corporations – fun and not measuring results. Let's take them one at a time.

First, let's talk about fun at work. There are two statements I detest. 'Business is business' and 'if it was fun they wouldn't call it work.' These are two of the most outrageous statements I have ever heard, but I keep hearing them over and over and over. Here is something to think about. Assume a person devotes 50 hours a week to work. This includes commuting, working, etc. You are probably saying, 'yeah right, only 50 hours.' Now let's say the person works 30 years, 50 weeks a year.

**30 years X 50 weeks a year X
50 hours a week = 75,000 hours!**

75,000 hours. Let's say that again – 75,000 hours. And this is probably a low estimate. You will spend more time at work than with your family. Whoa! The scale is way out of balance. Do you know what the crazy thing is? We do it to ourselves. Because of phrases like 'if it was fun they wouldn't call it work' people just accept the fact that work should be some type of painful existence. This is crazy! I am not going to go into much depth with this point due to the fact there are countless books on the

"If it was fun, they wouldn't call it work."

WHAT the heck is that all about!?

subject. What I will do however is talk about how this relates to innovation and several of the topics we have discussed.

Action Item #22

Fill in the blanks...

A. I work this many hours per week _____

B. I plan to work this many weeks per year _____

C. I plan to work this many years _____

Lines A x B x C = _____

On a Scale of 1 to 10 (10 is highest) I have how much fun at work _____

Do I need to think about adding more fun to my work environment? Yes or No _____

A boss once told me that 'I was laughing too much at work.' Let's say this one more time. I was told I was laughing too much at work. This is one of the most absurd statements EVER! I was a Project Manager, not a funeral home director.

Let's go down the checklist of the Frustrating Five and see how this relates...

1. Fear of Change
The environment was stuffy and 'corporate' – I was different and was changing things in a way by introducing laughter.

2. Fear if not having Control
The boss needed to control things and I was not playing along.

3. Fear of public opinion
What would people think about my boss if he had employees that were laughing 'too much?'

4. The Need to Sound Important
Work is serious stuff, right? It is very important, right? Stop laughing!

5. It is easier to criticize and critique than to construct and create
It was just so much easier for my boss to keep us all stifled and constrained rather that to be open to the idea of introducing some laughter at work.

Now I know you may be tempted to think 'well, the problem with what your saying is...' blah, blah, blah. Please trust me. I was not being disrespectful. Looking back, I should have realized that if I wanted to work within that department I needed to conform to the

111

culture. This is a good lesson as to why the concepts in this book are most effective with an entire team or company, not with just one individual trying to change a whole group. With that said, however, a person can still change his or her behaviors and possibly become a model and inspiration for the rest of the team.

Are you ready to be that person?

On a side note, did you know that kids laugh about 75 times a day on average, but adults laugh only about 10 times a day? And most of the adults I talk to say, 'and 10 is on a good day.' What's up with that!

Story: Bad Movie Friday

I had a job where things were becoming very tense and it seemed like a little fun would be a good thing. A few friends and I came up with an idea we called Bad Movie Friday. We would pick a movie that was so bad that it was 'good' (Convoy with Chris Kristofferson and Ali McGraw was our first choice). We would bring in pizzas, soft drinks, and some dumb prizes and show the movie over lunch. And since the movie was so bad, everyone could talk during the movie, laugh, and have fun. Simple, inexpensive, and not crossing any lines. I was pumped!

I started to talk up the idea and created some small posters to promote the event. A few days later a good friend of mine came up to me and said, "Rich, people are talking behind your back. They are saying you don't have enough to do." Fun was perceived as 'not busy.'

I hope one of the pieces of data you are taking away from reading this book is that I like having fun. I have learned from my 15 years in corporate America that fun can lead to problems unless, and this is the key point, *it is part of the culture and people know it is one of the things that is valued.* A POP! Culture values fun.

I often use this analogy to show just how important fun is in a work environment. Imagine I said to you, "We are going to go see a movie. The movie is three and a half hours." What is your first reaction? Most people, including myself, would think, 'wow, that seems really long for a movie.' Millions of dollars have been spent to *entertain* you and the reaction is hesitation. So what do corporations do? They have eight hour training courses with little to no fun activities and wonder why people walk away exhausted from boredom. My POP! Culture workshops are two hours, give or take 30 minutes depending on the size of the group. The sessions are so energetic and participatory that I rarely need to schedule a break. My view is that the workshop needs to be educational, provide tools people can use immediately, help teams and companies work on real life situations, AND be a fun experience. In my approach fun and work co-exist. They are not mutually exclusive. Asking my participants to moo like cows leads to phenomenal learning points and in depth discussions. Remember, we are smashing paradigms.

Question: If people have trouble sitting through a 3 hour movie, how can you expect them to sit through a 3 hour meeting?

The company where my friends and I proposed Bad Movie Friday was the same organization where a senior manager had been told by our CEO to 'come up with some creative ways to motivate our sales people and distinguish us from our competitors.' Here's what happened…

Story: Hair for Share

A few of us were sitting around on a Friday afternoon talking with a senior manager throwing out some fun ideas to increase sales. The manager was someone who had a good sense of humor, was a nice person, and was known as a 'pretty flashy dresser.' In fact, his hair was always combed back perfectly to the extent he looked like he had just come out of a hair salon. I threw out an idea. "Let's do a campaign called hair for share," I said. Everyone looked puzzled. "Let's introduce a campaign to the sales force that rewards the person who raises market share the most in his or her territory with a trip to the home office to publicly shave the manager's head. We could video tape it and even have some posters." The room went crazy! A burst of energy engulfed us. Some folks started with the typical 'but how will we know how to measure it blah blah blah' while other folks started to get really excited. Even the manager! This was huge! But guess what happened? Someone got the marketing department involved, big discussions took place, red tape emerged, and poof, the idea was killed by the typical corporate garbage that causes innovation to fail. Instead we gave out more donuts and pizzas. Wow, that must have blown customers away…NOT!

So what is the lesson here? Having fun is not a distraction! I always felt bad for the 'dot com' companies who brought in pool tables and dartboards, only to have them repossessed when the NASDAQ bubble burst. I actually heard some people say, 'see, that stuff doesn't work,' as if to say that having fun was the reason these companies went out of business. The critics seemed happy that the typical stifling, non-creative, low fun corporate culture had 'won out' in the long run. Back to suits and ties! My opinion is that pool tables and dartboards don't define a company's culture. They are merely toys to play with. Pool tables and dartboards don't equal innovation and creativity any more than suits and ties do. It is behaviors, attitudes, and ultimately respect and openness that define an innovative and creative culture. Remember the example about Apollo 13 and the engineers and artists earlier in the book? The engineers all wore ties.

Think about it this way. You can live in the nicest neighborhood in town in the nicest house, but if your family life is chaos, you live in the *worst home* on the block. The same is true for companies. Just because you wheel in a pool table next week and demand people start having fun – or else! – it doesn't mean you have a fun, open, and respectful corporate culture. You have to live it. The behaviors have to model it. You have to sincerely want to have fun, not have 'forced fun.' It doesn't have to be pool tables, Hair for Share, or Bad Movie Friday. Just have fun in a way that works for your company that will help promote and encourage the

115

behaviors and mindset of a POP! Culture and the NEWIDEA! process.

If you're not having fun, then implementing a POP! Culture and the NEWIDEA! process will be 'just another initiative.' If I gave you the choice of having fun or not having fun, who in their right mind says, 'you know what, I think I want to have a really miserable day today. Let's not have any fun.' It is the awful phrases like 'if it was fun, they wouldn't call it work' that are creating barriers to becoming innovative. Have fun with this. Get people out of their comfort zones. As I mentioned before, in my workshops I have people use squeaky hammers, wear beanie caps with propellers, make barn yard animal noises and laugh as much as possible. The goal is to make people feel safe and show them that it is okay to be a little 'wacky' at times. This helps lead to the idea explosion we have talked about.

As for the phrase 'business is business,' this goes completely against the concept of a POP! Culture. Business is NOT just business. Business is people working with people. Business = people. Unfortunately business often becomes about winning at all costs, even at the expense of others. A good friend of mine was giving me some feedback on a draft of this book and he said, "some people may read this book and think that a person should just get over it if he or she feels *devastated*, *crushed*, or *upset* when an idea is shot down. People may say this is just business." No! Business is not just business. Just because you have a title, wear a tie, or have a fancy business card, you do not have the right to be disrespectful. Healthy debate and constructive criticism are needed. However, insulting and disrespectful behavior can't be accepted if you sincerely want the ideas and input of others. I often ask people, "do you lead by fear or do you lead by inspiration?" Fear kills ideas. Inspiration leads to idea explosions. Lead by inspiration.

Try not to measure

This brings me to my point about not measuring the results of my approach to innovation. I do agree that measurements can be helpful to determine a return on investment and to see if 'something is working.' What I would like to propose is that you have trust that if people are promoting behaviors that model openness and respect and they are following the NEWIDEA! process, great things will naturally happen. Why complicate things with spreadsheets, reports, and meetings?

My wife and I sent our daughter to a pre-school when she was about two years old. There were no tests. There were no grades. How would I know if the school was 'working?' Does she share? Does she have a great imagination? Does she try new things? Does she say please and thank you? Can you imagine if I went in and said, "I have some baseline data of how often my daughter says 'thank you' and 'please' and how often she shares with other kids. I have been tracking this and have this fancy PowerPoint graph that illustrates her sharing tendencies. Could you send me a report each week and point out any trends or significant data points? Also, how about we meet every week to review the data? We could have fancy memos and meeting requests to make it look important."

Sound crazy? Then why do it inside a company! Have faith. Have trust. Measure it if you must, but know that all you will be doing is taking time and resources away from doing things like working with clients and leading your employees. This is where you will see the results. By following the simple approach in this book you will not need to measure openness and respect, you will simply have to look at one number – your bottom line. I tell clients I do have a process to measure results, but it is usually so a person can 'sell' the investment of time and resources needed to create a POP! Culture to others in their organization. If that is what you need to get your company to become innovative, then by all means do it. Again, do whatever works for *you* and *your* team or company.

Story: The Red Stickers
Here is an example of where measurements came into play. While conducting a workshop with a group of senior managers at a large wholesale and retail goods supplier, we had come to the portion of the session where the group was brainstorming ways to model the NEWIDEA! process. One person said, "How about for N – No negativity, we give people who squash people's ideas a red sticker. At the end of the year the person with the most red stickers doesn't get a bonus." Boom! You

would have thought by the reaction in the room that he had suggested that everyone cut off their toes. People were saying 'no way,' 'yeah right,' and even 'you are out of your mind.' It was as if the first half of the workshop had never happened and the NEWIDEA! process didn't exist.

I had to jump in and say, "whoa, whoa, whoa! Hold on!" It actually took a few minutes to calm the group down and make sure the person recovered from the 'attack.' I said, "let's try out the NEWIDEA! process and see what happens. John, what is your idea?" He said it again. We all practiced N – No Negativity. This time there were no outrageous reactions. I used a method/phrase they had developed to E – Encourage him to continue and we all waited and listened (W – Wait and Listen). This is what he said. "This is a company where we measure everything. It's just the way it is. Also, we live by the saying 'the things that get measured are the things that get done.' My thought with the red stickers is that it is a way to measure this initiative and to get everyone bought into the concept."

Then something amazing happened. A person asked, "can I do 'I,' include input?" We all said sure. The person said, "how about we build on John's idea and we flip it around. How about we give out gold stickers to people who encourage respect and openness and give the person who has the most gold stickers at the end of  the year double their bonus?" The group exploded in delight! We hadn't shot down John's idea. Instead we had listened, understood, and then built upon it. John's idea wasn't so much about red stickers and taking away bonuses as it was about measurements, recognition, and BUY-IN! Plus the people in the room would not be at risk for losing their bonuses because they were not open or respectful. [Hmmm, this was with senior management. Maybe we should have explored that further.] We then went through the rest of the NEWIDEA! process and came up with a plan that worked for them. There are many lessons but with respect to measuring results this was clearly a culture where it was needed – and that is fine!

Here are some ideas to help you measure the results of implementing a POP! Culture and the NEWIDEA! process:
1. Let your employees help determine the measurements.
2. Let your employees help determine the measurements.
3. Let your employees help determine the measurements.

What am I trying to say here? Rather than YOU telling people what will be measured, it may be worth your while to let the people who are being measured have a say in the process. The buy-in will be tremendously higher – as shown in the previous example.

In addition, it may be worthwhile to conduct some simple anonymous online surveys to establish a baseline of just how open your employees and your culture are to new ideas. Here are some possible questions to ask:

1. When I offer a suggestion or idea at work the response is encouraging. 1 would be never, 5 would be always.

2. I often hear the phrase "the problem with that is" when an idea or suggestion is presented at work. 1 would be never, 5 would be always.

3. People in many of the meetings I attend initially respond to an idea with the reason the ideas "won't work." 1 would be never, 5 would be always.

4. The department I work in is open to new ideas. 1 is not open at all, 5 very open.

5. The company I work for is open to new ideas. 1 is not open at all, 5 very open.

If you would like to see an example of an online survey that can be customized please visit www.innovationiseasy.com

The idea with these questions is to have a very simple approach to a measurement process. But what are we measuring? Respect and openness. That is the key to innovation. The danger is trying to measure pure dollars and cents. It can be done, but I would propose that a dollar figure is not the right thing to measure. It may, however, be helpful in your culture to measure the responses to questions like the ones above. Then, once you have this data, you can easily determine if your environment is fostering ideas and from there identify some tangible successes since the initiative has been implemented.

As I stated in the beginning of this book, the goal is to make innovation simple. Not to get bogged down in charts and graphs. In addition, the goal is to improve your bottom line by having everyone constantly sharing and implementing new ideas.

To recap, have fun, and if you can avoid it, don't measure this process. Your people will surprise you by how much they embrace the approach and the culture you are creating. Simply look at your bottom line.

I purposely have not asked my clients to *measure* their results. This is an experiment to test my hypothesis that a POP! Culture leads to bottom line results that are clear to everyone without having to add administrative tasks that take your focus away from customers and business objectives. Why create one more chart? With that said however, I can point to some results that have emerged from my efforts since the start of pursuing my mission:

1. A group of 75 accountants and finance managers at a Fortune 500 company have, as the client stated, "realized they are creative, and are generating and implementing great, new ideas on how to better do their jobs and meet the needs of their clients."

2. A team of restaurant general managers at one of the largest food service providers in America indicated that introducing the POP! Culture workshop at the beginning of an off-site meeting led to tangible actions the team could implement, specifically around an area of concern regarding a process that greatly impacted customers.

3. Over 100 lawyers working for a multi-billion dollar pharmaceutical company in three different office locations created and implemented a plan to better share information and enhance relationships among the offices to better meet the needs of their internal clients.

4. Team members at a multi-billion dollar insurance and financial services company reported that morale and productivity have greatly increased and that many of the tools taught in the POP! Culture workshop are *used every day* – 6 months after the workshops.

5. A prominent utility company initially intended to use the POP! Culture approach with its senior management team. However, due to the impact of the initial efforts, they have expanded the scope of the initiative to include field managers as well. In addition, the client conducted surveys of participants to collect baseline data that will be re-examined a short time after the first workshops are complete.

6. A leader at Fortune 20 company told me that the approach I shared with his team was outstanding and much easier and better than his company's "polluted" process around innovation.

7. A senior manager at a multi-billion auto parts distributor said, "I want to do NEWIDEA! However, I am concerned this could be so powerful we could have an idea revolution."

8. A team of participants at a well known news paper indicated that they "began using what they learned immediately" and that the workshop was "one of the best learning experiences they had ever had." Many are still using what they learned over one year after the workshop.

How about your company? Are you ready to see the bottom line impact of a POP! Culture in your organization. I am so confident that the approach I outline in this book works I tell my clients that if they are not 100% satisfied with our efforts, I tell them not to pay me. I have not had anyone take me up on that offer.

This is my mission. This is my passion. I believe in it. Have I inspired you enough to join me on this journey? There's more.

Chapter Three

Some things to
think about...

"We don't like their sound, and guitar music is on
the way out."
-- Decca Recording Co. rejecting the Beatles, 1962.

N ow that you have the basics of how to get started, let's look at a couple of key things to consider as you begin to implement the ideas and processes in this book.

1. Manage Expectations
2. Start Small and Prioritize
3. Follow Through

1. Manage Expectations

Creating a POP! Culture is ultimately about examining your work environment and establishing habits within individuals. That takes time. Did you know that it takes the average adult six to eight weeks to incorporate a new skill or discipline into his or her routine? And if you have a person who has felt a lack of trust or safety for a long period of time (possibly since they were six years old!), you'll need to be patient.

It is also important to stress the fun aspects of this approach and look into the possibility of not putting measurements around the process. Remember, the goal is to get everyone open to ideas and respecting others for their input – not to fill in fancy spreadsheets with data (unless you have to – sigh).

Finally, always be aware that people are different and they all have different ways of processing and sharing information. Shy and introverted people will still be shy and introverted (maybe) and extroverted people will still be extroverted (maybe). Each will still process and share data in a different manner. However, the shy person will feel safe to speak up and the extroverted person will be careful not to 'shoot from the hip' when ideas are presented, thus leading to more respect for others' input. It may be worth your while to invest in some good personality and temperament assessments when you start this process. Knowing how person X processes ideas and data compared to person Y can yield great dividends. I am purposely not going to spend a lot of time on this subject in this book since there is so much good work available on this topic (much of it is free and online!)

A friend shared something with me that brought this point home. She and each member of her team had recently completed a well-known personality assessment and to my surprise, she was described as 'an introvert.' She is one of the most outgoing people I have ever met! I asked her how could this be and she said, "I am an extrovert at home but more introverted at work. I am very cautious about sharing ideas or speaking up because I have seen what can happen." Please people, help me get this approach in wide spread use as quickly as possible! I can't take many more stories like this!

> Question:
> Do you know each team member's personality type and temperament?

2. Start Small and Prioritize

Once people feel safe and begin to see that yes, they are truly creative, you may have an explosion of ideas. This is a good thing! With this in mind, take a few 'easy ideas,' implement them, and celebrate the accomplishment. You will build momentum and increase credibility in your work environment. I always get a kick out of statements like 'we will have too many ideas.' What is really being said here? Is it a concern about time? We have already addressed that. Is it about control? Is it about not having all the answers and realizing others have great input? Is it concern about 'the kooks?'

You may ask, which ones are the 'easy ideas' and how do I prioritize them?

On the next page is a matrix you may have seen that can help prioritize your ideas. It is a simple four-quadrant approach that plots the potential impact of an idea versus the difficulty of implementation. This is designed to be an art, not a science, and yes there will be some discussions. It is not clear-cut. It is however, a tool that is simple and easy to use.

Which ones are the 'easy ideas' and how do I prioritize them?

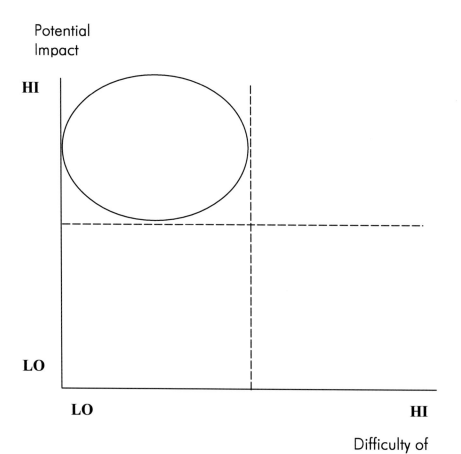

Potential
Impact

HI

LO

LO   HI

Difficulty of
Implementation

For example, I was conducting a workshop with a group that was brainstorming how to increase sales. We had each participant write ideas on sticky notes and then we drew the matrix I just described on a flipchart. People were free to write as many ideas as they wished and all ideas were then randomly placed on the flipchart without any judgment.

Potential
Impact

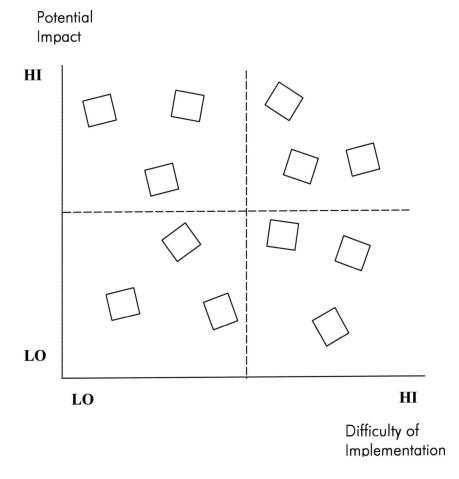

Difficulty of
Implementation

I then asked the participants to come forward to the flipchart, and place the sticky notes on the matrix based on potential impact and difficulty to implement. We also discarded any duplicates and clarified certain ideas – making sure we understood them – not judged them! We then decided what we would focus on first, using ideas on the sticky notes that were in the High Impact, Low Difficulty quadrant.

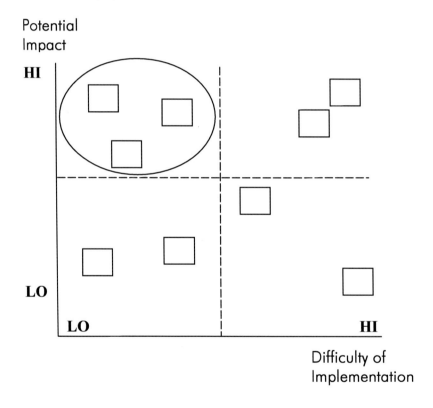

What are the big lessons here?

- o   Who brainstormed the ideas? The participants.
- o   Who plotted the sticky notes? The participants.
- o   Who decided what ideas would be worked on first? The participants.

Do you see a trend here? What do you think the level of buy-in will be?   How about any resistance to change? How passionate will people be since they are the ones who came up with the ideas, physically moved the sticky notes, and came up with the priorities?

Also, how about emotions in this process?   They are minimized. Why? Because each idea now belongs to the group, not to one person. This is the 'Mi idea es su idea' concept I described earlier. It is the *group* that is determining the impact and difficulty, not one or two people lobbying for their idea. In addition, by doing this prioritization exercise every idea is acknowledged and recognized.

Sticky notes are a very inexpensive and very powerful tool for idea generation. A few bucks on sticky notes in combination with a flipchart can go a long way – if the environment and culture is there to support them!

Action Item #23	Buy sticky notes and keep them in your meeting rooms, cubes, and offices – right next to the flipcharts!
	I will do this by when: _____

3. Follow Through

The final point I would like to make is the one thing you can do ensure your effort will succeed – follow through on your commitments. This sounds so basic but needs to be repeated over and over. If your mission is to have everyone open to ideas and respecting others for their input, guess what you need to do? Be open to ideas and respect others for their input! If you don't do it, how can you expect others to do so?

Story: Throw it to the right

Here is a story that is very touching and shows the power of being open to ideas AND following through. I saw this on ESPN and read about it on ESPN.com.

The head coach of the Notre Dame football team, Charlie Weis, had visited a very ill 10 year-old child days before a crucial game. In speaking with the child, he asked if there was anything he could do for him and the child gave the coach an idea for the first play of the game. He told Weis to "throw it to the right." The day before the game, the boy passed away. The coach told the team the story and had already talked with his quarterback about the first play.

As fate would have it, Notre Dame's first play came after they recovered a fumble and were backed all the way to their own 1 yard line – 99 yards from the other end zone. Logic and the most conservative play calling would result in a hand off and run to try and gain a couple of yards to create a little space for the offense to work. But what about the child's idea? Quarterback Brady Quinn asked the coach, "What are we going to do?" Weis said. "We have no choice. We're throwing it to the right." The play worked and Notre Dame gained 13 yards.

It broke with conventional wisdom. It may have been considered a 'kooky' idea. Can you imagine what would have happened if the play had backfired and none of us knew this story? "What a dumb play!" "What a bad idea!" "That's stupid!" All these statements and judgments would have been made – without ever understanding. Openness and respect. Openness and respect. The coach was open to an idea. He respected the child. Even if the play had 'not worked,' the coach would know inside he had done what was right. But it did work. He let go. He gave up control. He listened. He cared. He followed through. He made the kid's idea the team's idea. Everyone was bought in. He did not fear public opinion. He did not feel the need to sound important and tell the kid 'why the idea wouldn't work.'

If the head coach of Notre Dame football can do this with so much pressure, so much scrutiny, and so much at stake in terms of money and the demands of leading one of the most storied college football teams in history, can we all do what he did? Yes.

Hopefully you are now at a point where you are ready to say, "I want to begin my journey towards a POP! Culture! I want my company, my team, and myself to be a model for this new and simple approach to innovation. I want to create an idea explosion and help my company reap the benefits. I really want this. I want to make this part of my personal mission. "

Here is what you need to do.

Chapter Four

What do I need to do?

"Everything that can be invented has been invented."
--Charles H. Duell, Commissioner,
  U.S. Office of Patents, 1899

I have good news and bad news. The bad news is this is not something that you can just *talk* about with your team or company. The good news is a POP! Culture is something that is easy to create.

To get this process going you most likely will need to 'sell' the idea of introducing the concepts, approach, and action items in this book to your co-workers, your team, and maybe even your company. So how do you do that? Here is a simple road map and an approach you can use to 'sell' the idea of implementing a POP! Culture and the NEWIDEA! process.

1. Complete the action items
This will give you a good sense of where you are with the approach outlined in this book. Many of the action items are designed to be done in collaboration with other people so don't feel like your answers have to be perfect. The goal is to get the ball moving and get some initial thoughts on paper.

2. Determine who needs to be involved in establishing a POP! Culture within your team or company.
As we have talked about throughout this entire book, going at this alone is not the best approach. Make a list of people who need to be involved to help you come up with your 'sales pitch' for why you want to implement a POP! Culture. Note: even if you are the boss and you can technically *tell* people what to do, please consider the points about buy-in that have been shared throughout this book.

Name   When I will talk with him/her

_____   _____

_____   _____

_____   _____

_____   _____

_____   _____

3. Create and write down a plan for your 'sales pitch.'
Here are some questions to guide you in preparing your sales pitch. You will also need to determine your audience. Are you selling up, down, sideways – these are all factors to consider. Many of your answers will come from the action items you have already completed.

Q: Why create a POP! Culture? Think benefits and real dollar figures.

A: _____

_____

_____

_____

Q: What if we don't create a POP! Culture?

A: _____
_____
_____
_____

Q: What will a POP! Culture look like?

A: _____
_____
_____
_____

Q: What is in if for THEM [the people who will be impacted]?

A: _____
_____
_____
_____

Q: How will it impact THEIR day-to-day jobs?

A: _____
_____
_____
_____

Q: What is your proposed plan? Think about workshops, brainstorming elements of the NEWIDEA! process, getting additional information, and ways to make this fun.

A: _____

_____

_____

_____

Q: When will we start this process? Set a time frame and write down dates and milestones.

A: _____

_____

_____

_____

NOTE: Contact me if you need help! This is my mission – I want to help you succeed!
www.innovationiseasy.com

A final thought

If you follow the approach in this book there is a high probability you will be well on your way to making great things happen at your company. You will be part of the mission. You will be part of something big.

And maybe this goes even beyond companies. How about if families had a POP! Culture? How about if politicians used the NEWIDEA! process? What if kids at an early age learned to be open to new ideas and respect others for their input rather than be jaded by our current society? College? High School? Elementary School? Would it ever be too early to learn what I am proposing? Is there a place in society where this is not applicable? There are a lot of problems in the world today. The good news is the world is filled with great people who have great ideas — we just need to get the ideas out, create a culture that is open to ideas and respects everyone for their input, and have a simple process to get ideas to action.

Final story: *The Itchy Shirt*

A good friend of mine told me that this book was similar to what he tells his children — don't be an itchy shirt. Now what the heck does that mean?

He tells his children that an itchy shirt is one that people just don't want to wear. It is uncomfortable. It's irritating. People don't like it. It may look good but it just isn't worth the aggravation to put it on. So what happens to this shirt? It just gets put back on the rack and no one wants to wear it. He tells his kids 'don't be an itchy shirt.' Don't be someone that people don't want to be around and, in his kids' case, someone the other kids don't want to play with. You may look good and have the best toys, but if you are an itchy shirt you will end up on the rack and no one will want to deal with you.

The reason I like this story is that it summarizes my mission so well:

> *To inspire individuals and organizations to create respectful, open, and innovative work environments that promote the creation of new ideas.*

"The problem with that is," "that won't work because" and other phrases like this are itchy shirts. My goal, and hopefully yours after reading this book, is to take action and replace the irritating – and frustrating – reactions from 'itchy shirts' with openness and respect. Then you will truly realize the potential of ALL ideas.

In the spirit of Dr. Seuss, here's a little something to send you on your way.

I am innovative.
I am innovative.
Innovative I am.

Innovative I am.
Innovative I am.
I do not like my ideas being shot down, understand?

You don't like them shot down at work?
You don't like them shot down by a jerk?

Not at work!
Not by a jerk!

How about in your house?
Or online with a mouse?

Not in my house!
Not with a computer mouse!

Innovative I am.
Innovative I am.
I do not like my ideas being shot down, understand?

Wait, in a meeting with just two?
How about in a meeting with just two?
These people like to shoot down ideas.
Sadly it's true!

They will shoot them down without care.
They will shoot them down anywhere.
They will do it to your face!
They will do it without grace!

But Innovative I am.
Innovative I am.
I do not like my ideas being shot down, understand?

So what shall we do?
Where shall we go?
I have a process for you!
And a mind-set you know!

It is a simple approach
One that you'll yell loudly 'oh yeah!'
An approach called POP! Culture
And a process called NEWIDEA!

We are all innovative and creative you see
So please hear me out, please hear my plea!

You can be innovative on a boat.
You can be innovative with a goat.
You can be innovative in a box.
You can be innovative with a fox.

Innovative I am.
Innovative I am.
I do not like my ideas being shot down, understand?

So away with you
You Frustrating Five
I have 5 other great steps
In which innovation will thrive!

Forget the control and the fear of a change,
Public opinion can cause such disdain!
And sounding important, who needs that not me
And being closed to ideas is not cool so you see.

But wait! These five brand new steps
They are simple to use
To promote innovation
Here are the don'ts and the do's

Start with a mission
And what you will value to start
This is about being open
And respecting a lot

What's next you ask
To make this great change
You must work to get everyone
On the same page!

Once that is done
You can then start to prepare
For a simple process
We call NEWIDEA!

But always remember
When there is work to be done
Always remember
To make the job fun!

And measure if you must
But do that with care
Because the easier it is
The more people will share

So go now and remember
Don't treat people with scorn
And don't ever say to someone
'Mustard doesn't go on corn!'

And guess what? Mustard really does go on corn! Here is a recipe I found on the Kraft Foods web site. Enjoy! The web address I found this at is http://www.kraftfoods.com/recipes/SaladsSideDishes/ VegetablesSideDishes/Mustard-GlazedCorn.html

2 Tbsp. GREY POUPON Dijon Mustard

2 Tbsp. chopped fresh parsley

1 Tbsp. lemon juice

2 cloves garlic, minced

4 fresh ears of corn, husks and silk removed

1 Tbsp. olive oil

PREHEAT grill on medium heat. Combine mustard, lemon juice, parsley and garlic; set aside.
BRUSH corn with oil.
GRILL corn 8 to 10 min. or until tender, turning occasionally and brushing with the mustard mixture.

Who knew?

Summary of Action Items

Worksheets and additional information are at
www.innovationiseasy.com/actionitems.htm

Action Item	Check box when done
Conduct the Balloon Effect™ with a colleague	
Tell a person his or her idea "won't work" and discuss how they felt	
Identify what YOU are doing to promote or discourage innovation	
Create a line item in your next budget titled 'cost of not being open to ideas'	
Ask yourself if you have ever ignored changing services such as phone or insurance even though it may cost you less money to do so [people fear change]	
Ask a co-worker to do something out of the ordinary and see his or her reaction [people fear public opinion]	
High five the next person you see and discuss the reaction	
Identify a person who is a Republican or one who is a Democrat. Go to that person and criticize something George W. Bush or Bill Clinton did, depending on your audience. [People want to win]	
Take a day and write down all of the non-productive activity you see in your company and log how much time is wasted. Decide if you have an extra two minutes to listen to 'kooks' and their ideas.	
Take a day to identify examples of the Frustrating Five™ and the impact they had	
Write down names, dates, and resources needed to complete your company, team, and individual innovation mission statements	
Write a draft of your innovation mission statement	

Action Item	Check box when done
Write down what you will value and what you will not tolerate in your POP! Culture™	
Answer this question – what if I have a person who has a great performance record and strong abilities but does not embrace our values?	
Answer the questions 'Why are we creating a POP! Culture?' and 'What if we don't?' Think ROI.	
Write down what is in it for your team or company if you create a POP! Culture?	
List out the questions and possible answers so people can know how a POP! Culture will affect their day-to-day jobs.	
Write out a 'picture' of what your team or company will look like once you have implemented a POP! Culture.	
Imagine I am sitting next to you as you are reading this book. I say, "excuse me, can I suggest an idea? I would like to suggest that we give all the power in this country to the Republican Party. Write down your first thought.	
Conduct a meeting with your team to brainstorm ways to model the behaviors in the NEWIDEA! process.	
Get a flipchart and markers for every conference room and office – and use them!	
Calculate how many hours you will work in your career and decide if you need to have more fun at work.	
Buy sticky notes and keep them in your meeting rooms, cubes, and offices – right next to the flipcharts!	
Complete the action plan to implement a POP! Culture in your organization.	

The Innovation Company, LLC

www.innovationiseasy.com
978-266-0012
info@innovationiseasy.com

Workshops
Retreats
Keynotes
Management consulting
Coaching

The Innovation Company's mission and vision

Mission
The mission of The Innovation Company is to inspire individuals and organizations to create respectful, open, and innovative work environments that encourage people to constantly share and implement new ideas.

Vision
The vision of The Innovation Company is one in which all individuals, regardless of their position in the workplace or society, can present new ideas and suggestions and receive a respectful response from others that demonstrates the sincere willingness to be open to different points of views.

About the author

Rich Trombetta is an 18-year veteran of corporate America, working in various roles for such organizations as GE Capital, Fidelity Investments and Thomson Financial. In addition, he worked as a sports producer for an ABC affiliate in Providence, Rhode Island, was an NBC Page in New York City.

He is currently the President of The Innovation Company, LLC located in Acton, Massachusetts. He works with companies that want to get EVERY employee constantly sharing and implementing new ideas. His clients include Pfizer Pharmaceuticals, Monster.com, , Children's Hospital of Boston, AIG, and CARQUEST.

Rich is involved with several non-profit organizations and even co-founded his own non-profit company, SpeakUp, Inc., which taught presentation skills to urban youths. He also serves on the Board of Directors of Generations Incorporated, a leader in uniting children and older individuals to improve literacy in Boston's public schools.

Rich holds a degree in Electrical Engineering from the University of Massachusetts at Amherst and has completed graduate work at Northeastern University. He lives in the Boston, Massachusetts area with his wife, two daughters, and two cats.

ISBN 141207999-3